EDUCATION IN A COMPETITIVE AND GLOBALIZING WORLD

S.T.E.M. EDUCATION

STRATEGIES FOR TEACHING LEARNERS WITH SPECIAL NEEDS

EDUCATION IN A COMPETITIVE AND GLOBALIZING WORLD

Additional books in this series can be found on Nova's website under the Series tab.

Additional e-books in this series can be found on Nova's website under the e-book tab.

EDUCATION IN A COMPETITIVE AND GLOBALIZING WORLD

S.T.E.M. EDUCATION

STRATEGIES FOR TEACHING LEARNERS WITH SPECIAL NEEDS

SATASHA L. GREEN
EDITOR

nova publishers
New York

Library of Congress Cataloging-in-Publication Data

ISBN: 978-1-63117-806-1

Published by Nova Science Publishers, Inc. † New York

CONTENTS

PREFACE

Advancing education in science, technology, engineering, and mathematics (STEM) in U.S. public schools has been at the forefront of educational issues and a national priority (President's Council of Advisors on Science and Technology, 2010). Although there is a need for this ambitious initiative students with disabilities has been left out of the conversation. Individuals with disabilities have been underrepresented in STEM fields for many years. Traditionally individuals with disabilities in STEM careers lag even further behind discrepancies of race and gender in these areas. Therefore, the need to provide general and special education teachers practices and strategies to improve outcomes for students with disabilities in STEM areas is imperative.

The nation's changing demographics and continued need to remain globally competitive makes it clear that general and special education teachers need strategies to support, instruct and engage student with disabilities in STEM education. Students in U.S. schools are academically behind their international peers in STEM areas. Currently, the United States ranks 17th in science and 25th in mathematics among other nations (National Center for Education Statistics, 2011). In the field of engineering, college programs in China and India graduated many more engineers than in the U.S. (Gerefii, Wadhwa, Rissing, & Ong, 2008). For example, in 2011, China's engineering graduates totaled one million (Shammas, 2011), as compared to colleges in the U.S. which graduated 84,599 engineers (Deffree, 2012).

President Obama stated that it is a "national imperative," to train 100,000 STEM college graduates over the next decade (America Chemical Society, 2012). These efforts are also aimed at attracting underrepresented groups such as girls and persons of color into the STEM pipeline (Custer & Daugherty, 2009) however individuals with disabilities have been left out of the equation. To accomplish President Obama's goals U.S. teachers and educational professionals must educate and engage all students, including students with disabilities to pursue STEM disciplines (Community for Advancing Discovery Research in Education, 2011).

There is universal agreement that teachers do matter and, moreover, there exists empirical support for the proposition that student learning is affected by the qualifications of teachers. This is especially true in mathematics, which is the foundation for all future STEM learning (Community for Advancing Discovery in Education, 2011). Although almost all U.S. teachers hold at least basic qualifications (e.g., a bachelor's degree and teaching certification), many are teaching subjects for which they lack adequate academic training, certification, or both. This becomes even more problematic when they have a student with a disability in their classroom. Ingersoll (1999, 2002, 2003) found that about a third of all secondary school teachers who teach mathematics do not have either a major or minor in math, math education, or related disciplines like engineering or physics. In science, about one fifth of all secondary school teachers do not have at least a minor in one of the sciences or in science education. The data clearly indicates that many U.S. students are taught by under-qualified math and science teachers in U.S. schools. Therefore, providing both general and special education teachers practices and strategies to improve outcomes for students with disabilities in STEM is needed.

Another area of major concern is the teaching of subject matter in STEM education, specifically the integration of technology into math and science concepts. Technology may not be infused into the curriculum and engineering in many cases is omitted or causes confusion in how it is related to science and mathematics curricula (Vest, 2009). As a result, very few K-12 teachers have adequate preparation to teach engineering concepts and content (Custer & Daugherty, 2009); this becomes even more problematic when teaching students with disabilities who may need differentiated instruction to meet their learning needs. According to the National Academy of Engineering and the National Research Council (2009), science and mathematics are typically taught in "silos," as separate, independent subjects. This teaching method can affect the quality of instruction in STEM which requires deep content knowledge (in all four areas) in addition to an expertise in teaching (Community for Advancing Discovery Research in Education, 2011). Therefore, it is imperative for general and special education teachers to have instructional strategies that help to improve the learning outcomes for their students with disabilities in STEM.

To address this pressing need to provide general and special education teachers practices and strategies to improve outcomes for students with disabilities in science, technology, engineering and mathematics, *STEM Education: Strategies for Teaching Learners with Special Needs* provides teachers and educational professionals the knowledge, skills, practices, and strategies to support learners with disabilities in STEM education. This book is intended for undergraduate and graduate students enrolled in methods courses in Colleges of Education, College of Arts and Sciences, and Institutes of Technology. More specifically, this book provides background information to prepare K-12 general and special education teachers and educational professionals in pedagogy for integrated inquiry-based teaching and learning for students with special needs in STEM concepts. This book will also help to provide: (a) ideas about adaptation to STEM content for learners with special needs to meet student learning outcomes and (b) general and special educators with the knowledge, skills and resources for effective STEM teaching and learning for students with special needs.

As noted earlier, the primary goal of this book is to provide K-12 general and special education teachers practices and strategies in STEM content to support the learning needs of their students with disabilities. Therefore, K-12 general and special education teachers and education professionals will (1) increase STEM content knowledge and understanding of authentic STEM applications for K-12 students with disabilities; (2) develop expertise in pedagogical approaches such as authentic and active project-based learning for students with special needs; (3) utilize strategies and resources to integrate technology into STEM teaching and learning for K-12 students with disabilities; and (4) increase their knowledge base, expertise, and experiences in differentiating STEM instruction from traditional instruction for learners with disabilities from culturally and linguistically diverse backgrounds.

The authors in this book address practices and strategies to successfully implement STEM education for students in specific disability categories. In Chapter 1, Davis discusses the huge push for Science, Technology, Engineering, and Mathematics (STEM) education in the United States. The push is supported by government and leading organizations due to the current global economy, international ranking of U.S. students in mathematics and science, and shortages of people pursuing STEM careers. She examines several strategies that can be used to increase student access to STEM content including, the use of UDL, technology, and accommodations. According to Davis, students with disabilities are not provided appropriate information or counseling related to taking advanced courses or STEM career preparation.

Obi, in Chapter 2, points out that for students with communication disorders (CD) to benefit from STEM, student and teacher collaboration must be an important component in any training the students receive in STEM. The author notes that teachers must also make a sincere effort to learn from speech and language professionals and vice-versa in order for students with CD to achieve success. It is obvious that enhancing STEM programs to benefit students with CD requires a process of system change, as opposed to isolated programs and invalidated instructional practices often common with programming for these students.

Obi, in Chapter 3, asserts scientific fields make increasing use of technology and STEM content, new opportunities emerge for students with a broad range of abilities. Moreover, when students with disabilities and STEM teachers form learning partnerships, the possibilities for academic and career success multiply. More specifically, when the student and teacher work together to develop creative alternatives (accommodations) for challenges faced by students with cognitive disabilities, their outcomes improve. This author discusses K-12 general and special education teacher practices and strategies in STEM content to support the learning needs of students with cognitive disabilities.

In Chapter 4, Sorrells, Cole, Pazey and Carter present research-based instructional frameworks and strategies that can be used in STEM classrooms. The authors provide an example of a successful school model that is helping students with disabilities excel in STEM subjects. Contemporary STEM initiatives and programs are aimed at increasing student achievement in math and science in order to prepare them for careers in technology based fields; however, not all students are showing gains in STEM subjects. Students with learning disabilities have yet to show significant achievement gains in math and science subjects, which is problematic considering that this population of students is an untapped talent pool that can contribute to the STEM workforce.

Gibson and Obiakor, in Chapter 5, discuss the impact that the difficulties faced by students with emotional/behavioral disorders (EBD) have on their ability to be successful in STEM education. Embedded in this discussion are some evidence based solutions to increase the likelihood that these students will be successful with STEM education. Considering that many school programs moving towards curriculum that emphasize STEM education, it is critical to discuss ways to support students with EBD, and in particular culturally and linguistically diverse students with EBD.

Van Bergeijk, Ranaldo and, Shtayermman, in Chapter 6, discuss the challenges in teaching students with Autism Spectrum Disorders (ASDs). Now, teachers of STEM are faced with the unique challenges and rewards of teaching this population. Teaching the core content of STEM will not be the challenge. In fact, these students will be enthusiastic learners of the STEM material and their depth of knowledge may even exceed that of the teacher's. Rather, the challenge will be managing the co-morbid disorders of ASDs and helping students with ASDs overcome their impairments in executive functioning, social skills development, and independent living skills. The authors assert that teachers of STEM will need to work with a variety of professional support staff such as school social workers, psychologists, speech and language therapists, and occupational therapists to address some of these issues.

Utley and Obiakor, in Chapter 7, describe causes, symptoms, and challenges following traumatic brain injury (TBI) (e.g., physical, emotional, and cognitive difficulties). They distinguish mild TBI from other mild categories of disability and identify classroom interventions and strategies. These authors identify resources and services for educators, professionals, and parents and note the focus should be on providing children and adolescents with TBI with opportunities to be placed in the general education classrooms at the earliest possible point.

Working with Learners with Physical and Health Impairments in STEM by Turton, in Chapter 8, addresses issues surrounding the increased access to STEM within the realms of special education and more specifically the category of physical and health impairments. This author examines approaches that are suitable to teach individuals with physical and health impairments and how the shifting landscape of STEM educational discourse can be grounded in this arena. All of this is surrounded by the frameworks of schooling characteristics, strategies and policy of STEM and access to its 'pipeline' for these students.

Eleweke, in Chapter 9, discusses the implications of utilizing educational materials and programs with rich visual effects such as videos, animations, 3-Ds, and subtitles; having effective teachers who plan and address the unique learning needs of learners with hearing loss; and supporting learners with hearing loss to be actively involved in their own learning process to facilitate the development of STEM knowledge and skills.

Edwards and Green, in Chapter 10, discuss the numbers of students with visual impairments entering into STEM-related courses in higher education and STEM fields. There has been a slow but steady increase. This increase can be credited to the attitudes, support and preparation of K-12 students in math and science. As mathematics and science at the K-12 level continue to serve as the "gatekeepers" to entry into STEM programs in higher education, teachers in K-12 classrooms are charged with the task of preparing students in these subjects, not only through traditional pedagogy, but through a clear understanding of the needs of all students.

Torres-Velasquez, Robert-Harris, Lopez-Leiva, Westby, Lobo, Dray, Martinez, Astigarraga and Aguilar-Valdez, in Chapter 11, describe the need to implement equitable, instructional approaches to differentiate STEM education for English Language Learners (ELLs) with special needs. Based on frameworks such as Ganma, Nepantla, and Multiliteracies, the authors suggest a model that they call, *Complex Integrated Curriculum.* This student-centered model presents a differentiated STEM curricular approach that emphasizes learning and owning of knowledge processes that are parallel to the development of students' STEM identities (as student and as scientists). This curricular model carefully considers a responsive and adequate differentiation of the content, the process or activities, and the products that support the teaching and learning of ELLs with disabilities in STEM fields.

Working with culturally and linguistically diverse (CLD) learners with special needs in STEM by Cooley-Nichols and Sheffield, in Chapter 12, discuss the learning styles of CLD students, the importance of STEM instruction for CLD students with special needs and instructional strategies that support STEM instruction for CLD students with special needs.

While STEM is not typically discussed as an approach to address the needs of CLD students with special needs, they are an ideal group to target for STEM instruction and careers.

The final chapter by Dunn, Rabren, Russell, Massey and Mairtin, Chapter 13, highlights strategies that promote and enhance the participation of persons with disabilities in STEM degree programs and careers. These authors discuss transitioning students with disabilities into STEM and provide an overview of the outcomes of adults with disabilities and the legal requirements related to the transition for youth with disabilities which provides the context for practices for transitioning learners with disabilities into STEM.

Collectively these chapters address a range of issues in K-12 and higher education that are important to working with learners with special needs in STEM education. In order to change the STEM landscape and to address President Obama's initiatives, efforts must be aimed at attracting underrepresented groups such as girls and persons of color into the STEM pipeline (Custer & Daugherty, 2009); however, individuals with disabilities have to be a part of the equation. Teachers and educational professionals must educate and engage *all* students, including students with disabilities to pursue STEM disciplines (Community for Advancing Discovery Research in Education, 2011).

ACKNOWLEDGMENTS

I gratefully acknowledge Dr. Cheryl A. Utley, Research Associate Professor at Chicago State University, a published author who served as the Guest Editor for this book. Dr. Utley has written several books, books chapters, peer-reviewed journal articles, and theory-based articles that focus on multicultural special education, intervention research and culturally responsive practices for students with and without disabilities. These published texts serve to facilitate the creation of inclusive classrooms that are accepting of all children regardless of their diverse needs. This philosophy is also illustrated in Dr. Utley's scholarly presentations at the state, national and international levels.

Dr. Utley serves on several editorial boards for peer-reviewed journals. Among her many awards are the Post-Doctoral Fellowship at Juniper Gardens Children's Project-University of Kansas, Who's Who Among Young American Professionals, Who's Who in American Education, Wisconsin Center for Research's Pre-doctoral Scholar, Marie Christine Kohler Fellow, and Advanced Opportunity Fellow. Dr. Utley provided important contributions to this book and essential research and editing support that helped to strengthen the book.

Additionally, my most important partners in this effort have been those who wrote chapters. Their expertise in special education, STEM Education, transition from school to career and higher education, and culturally responsive teaching each bring a piece of the puzzle to help create a complete picture for working with learners with special needs in STEM Education.

References

Community for Advancing Discovery Research in Education. (2011). Retrieved from http://cadrek12.org/projects/community-advancing-discovery-research-education-cadre-0.

Custer, R. L., & Daugherty, J. L. (2009). Professional development for teachers of engineering: Research and related activities. *The Bridge: K-12 Engineering Education 3*(39).

Deffree, S. (2012). *Engineering the Next Generation of STEM*. Retrieved from http://www.edn.com/electronics-blogs/other/4369012/Engineering-the-next-generation-of-STEM.

Gerefii, G. Wadhwa, V., Rissing, B., & Ong, R. (2008). Getting the numbers right: International engineering education in the United States, China, and India. *Journal of Engineering Education, 97*(1), 13-25.

Ingersoll, R. M. (1999). The problem of underqualified teachers in American secondary schools. *Educational Researcher, 28*(2), 26-37.

Ingersoll, R. M. (2002). *Out-of-field teaching, educational inequality, and the organization of schools: An exploratory analysis*. Seattle: University of Washington, Center for the Study of Teaching and Policy.

Ingersoll, R. M. (2003). *Out-of-field teaching and the limits of teacher policy.* Seattle: University of Washington, Center for the Study of Teaching and Policy.

National Center for Education Statistics. (2011). *Digest of education statistics,* Retrieved from http://nces.ed.gov/pubs2012/2012001.pdf.

National Academy of Engineering and National Research Council. (2009). *Engineering in K–12 education: Understanding the status and improving the prospects*, Washington, D.C: The National Academies Press.

National Research Council. (2012). *A framework for k-12 science education: Practices, crosscutting concepts, and core ideas.* National Academies Press. Washington, D.C.

President Obama's Council on Jobs and Competitiveness. (2012). Retrieved from: http://www.whitehouse.gov/administration/advisory-boards/jobs-council.

Shammas, M. (2001). *Number of US engineers in decline relative to China, India.* Retrieved from http://www.dukechronicle.com/article/number-us-engineers-decline-relative-china-india.

Vest, C. M. (2009). Putting the "E" in STEM education. *The Bridge: Linking engineering and society (3)*39, 3-4.

In: S.T.E.M. Education
Editor: Satasha L. Green

ISBN: 978-1-63117-806-1

Chapter 1

THE NEED FOR STEM EDUCATION IN SPECIAL EDUCATION CURRICULUM AND INSTRUCTION

Kimberly E. Bryant Davis*
Oklahoma State University, US

ABSTRACT

There is a huge push for science, technology, engineering, and mathematics (STEM) education in the United States. The push is supported by government and leading organizations due to the current global economy, international ranking of U.S. students in mathematics and science, and shortages of people pursuing STEM careers. One solution for alleviating potential shortages is to include people with disabilities in STEM education. There are barriers for students with disabilities in pursuing STEM careers including lack of role models and mentors, appropriate instruction, low expectations and lack of encouragement from influential adults. There are several strategies that can be used to increase student access to STEM content including, the use of UDL, technology, and accommodations. Advantages with students with disabilities include increased self-confidence, self-advocacy and self-determination skills.

INTRODUCTION

The current society is a technological, innovative culture. Friedman (2006) expressed the current state of affairs when he declared the world flat. No longer are students solely competing with their American peers. The current society is a global one. Advances in technology have played a key role in the transformation to a global environment. Science, technology, engineering, and mathematics (STEM) have largely influenced the advances that affected global leadership and advanced the global economy (Dunn, Rabren, Taylor, & Dotson, 2012).

* Email: kimbeed@okstate.edu

The resulting global environment results in a global economy. The current global economy has led to an increase in the need for workers in the fields of STEM (Alston & Hampton, 2000; National Science Board, 2006; Tyson, Lee, Borman, & Hanson, 2007). STEM careers typically require postsecondary preparation. As a result STEM education has become a priority for the United States (Burgstahler & Chang, 2009; Tyson, et al., 2007).

With the increased national emphasis on STEM professionals and the postsecondary education typically required for these professions where does this leave students with disabilities? The number of students with disabilities participating in postsecondary education is increasing (Burgstahler & Chang, 2009; Burgstahler & Doyle, 2005; Henderson, 1999, 2001). With the requirement that most STEM careers require some postsecondary preparation the increase in the number of students with disabilities attending postsecondary education is a step in the right direction; however, when compared to their nondisabled peers fewer students with disabilities graduate from high school and enroll in institutions of higher education. Of the students with disabilities who are able to reach these milestones, fewer earn a degree or certificate (Burgstahler & Doyle, 2005). The current status indicates few students with disabilities are pursuing STEM careers and the attrition rate of those students who do is high (Burgstahler & Doyle, 2005; National Science Foundation, 2000; Office of Disability Employment Policy, 2001); however, underrepresented populations, including students with disabilities, are being looked at as one way to meet the need for STEM professionals. Interest in assisting students with disabilities, to pursue STEM careers is increasing (Lam, Doverspike, Zhao, Zhe, & Menzemer, 2008; National Science Foundation, 2000). This chapter will introduce the idea of STEM and STEM education. The importance of STEM in K-12 settings will be addressed followed by the need for STEM education in special education curriculum and instruction. Finally, the chapter will address the benefits of STEM education in special education curriculum and instruction.

What Is STEM Education?

There is a huge push for STEM education in the United States. STEM education is the preparation for STEM fields and encouragement of STEM literacy. A commonly accepted definition of STEM education is provided by Tsupros, Kohler, and Hallinen (2009):

> STEM education is an interdisciplinary approach to learning where rigorous academic concepts are coupled with real-world lessons as students apply science, technology, engineering, and mathematics in contexts that make connections between school, community, work, and the global enterprise enabling the development of STEM literacy and with it the ability to compete in the new economy.

Within industry and education, STEM education has multiple interpretations. Often, STEM education represents four intertwined but separate disciplines (Basham, Israel, & Maynard, 2010). However, Bashma and colleagues (2010) cautioned that the separation of content areas is generic and continues to support practices that initiated our current difficulties in maintaining global leadership. More ideal would be an integrated approach to STEM education that is emphasized in the literature and demanded by the current global society. As Tsupros and colleagues (2009) mentioned in the definition the approach should be interdisciplinary.

STEM education is still a developing area. Ideas of how STEM education should be presented are varied. Lantz (2009) suggested seven ideas of what STEM education should represent. According to Lantz (2009), STEM education should:

1. Be trans-disciplinary in its overall approach;
2. Be driven by standards that complement the trans-disciplinary philosophy;
3. Use the backward mapping techniques advocated in Understanding by Design;
4. Use both problem-based and performance-based teaching and learning;
5. Use the 5E teaching and learning cycle to plan units and activities within the curriculum;
6. Be digital in format and coupled with digital teaching technologies such as whiteboards, tablets, student response systems, etc.; and
7. Use both formative and summative assessments with task and non-task specific rubrics. (p.8).

In addition to the ideas espoused by Lantz (2009), Morrison (2006) presented ideas related to STEM education (Marino & Beecher, 2010). Morrison (2006) suggested that STEM educated students should be:

- Problem solvers, able to frame problems as puzzles and then able to apply understanding and learning to these novel situations (argument and evidence).
- Innovators with "power to pursue independent and original investigation"(Gilman, 1898) using the design process.
- Inventors, recognizing the needs of the world and creatively designing and implementing solutions.
- Self-reliant, able to set their own agendas, develop and gain self-confidence and work within specified time frames.
- Logical thinkers- using the logic offered by calculus and found in 60% of all professions world-wide; able to make the kids of connections to affect an understanding of natural phenomena.
- Technologically literate, understanding the nature of the technology, mastering the skills needed and applying it appropriately (Knowledge, Ways of Thinking and Acting, and Capabilities as specified by ITEA in Technically Speaking).
- Participants in the STEM lexicon that supports the bridge between STEM education in school and the workplace.

Students should be able to relate their own culture and history to their education (p. 2-3). The National Science Teachers Association (NSTA) further evidences the importance of STEM education and the Next Generation Science Standards (NGSS), which suggest nationwide STEM, focused standards. The standards are steeped in science; however, they are heavily influenced by engineering. The NGSS "establishes learning expectations for students that integrate three important dimensions: (a) science and engineering practices, (b) disciplinary core ideas, and (c) crosscutting concepts; effectively builds science concepts from K-12 and integrates concepts of engineering (National Science Teachers Association, 2013)". The Common Core puts forth standards related to mathematics which are supported by the National Council of Teachers of Mathematics (NCTM). These standards stress the importance of processes and proficiencies and encourage the use of inquiry within mathematics. The standards support STEM to a lesser extent than the NGSS, but still demonstrates support of STEM education (National Governors Association & Council of Chief State School Officers, 2013). The goal of STEM education is to provide integrated content instruction. Instruction should be deliberate in teaching concepts related to each content area through integration, weaving in and out of content areas almost seamlessly. The major objective is to move away from prior practices of teaching content in silos, separate and discriminate from other content areas. (See Table 1).

Table 1. Common Misconceptions of STEM Education

Technology means additional computers and hardware for schools and students	False. Sometimes the best technology for instruction includes low-tech/no-tech solutions.
STEM Education consists only of science and mathematics.	False. Technology and Engineering play vital roles in STEM Education.
STEM Education addresses workforce issues.	False. STEM is important for those who will pursue STEM careers, but STEM literate citizens are better prepared for today's society.
Mathematics Education is not part of science education.	False. There are often overlap between the subjects with opportunities to discuss the content areas collectively.

IMPORTANCE OF STEM IN K-12 SETTINGS

Global Leadership

There is good reason for the increased emphasis on STEM education. In 2005 the National Academies Press released a report entitled Rising Above the Gathering Storm: Energizing and Employing America for a Brighter Future (Committee on Science, 2005). This report emphasized the need for STEM education in order to secure the United States as a leader and competitor in a global market. The report put forth four recommendations:

1) Increase America's talent pool by vastly improving K-12 mathematics and science education;
2) Sustain and strengthen our nation's commitment to long-term basic research;
3) Develop, recruit, and retain top students, scientists, and engineers from both the United States and abroad; and
4) Ensure that the United States is the premier place in the world for innovation.

This report was not the only indication that the United States may need to reassesses the current educational environment. Other reports such as The *Trends in International Mathematics and Science Study* (TIMSS) and *The Program for International Student Assessment* (PISA) also indicated the United States lagging behind other countries in mathematics and science. The 2011 TIMSS revealed US ranking 11th in mathematics overall and 7th in science overall (Provasnik et al., 2012). The 2009 PISA indicated US students scored below average, ranking 25th overall in mathematics literacy and 17th overall in science literacy (Fleischman, Hopstock, Pelczar, Shelley, & Xie, 2010). Even though there are sampling issues that have been raised as to international participation, some would consider this an indication that the United States is losing ground internationally.

There are some factors that have affected international standings in mathematics and science. Lantz (2009) identified several factors. One factor, is that secondary schools continue to be departmentalized and even though STEM demands content to be integrated. Content continues to be taught in isolation. A second factor, is elementary teachers typically teach all content areas, which provides a better opportunity for integration, however, many elementary teachers are ill-prepared to teach mathematics and science effectively (Lantz, 2009). The result means students enter secondary education without solid foundations in mathematics or science. These results, coupled with the workforce shortages in STEM careers concern many and have led a push by the government and other leading organizations for STEM education.

Post-secondary Preparation

Maintaining status as a global leader is not the only reason for emphasizing STEM education. There are STEM career shortages, but students also need to be prepared to work and live independently in an innovative, technological society. For students who are not preparing for STEM careers STEM literacy is still emphasized. Preparing students to be STEM literate will assist with increasing their post school opportunities. To assist with STEM literacy students need access to STEM content, opportunities to learn STEM content and meaningful connections to STEM content (Basham, et al., 2010). Preparing STEM literate students has the potential to alleviate shortages in STEM careers in addition to preparing students for an increasingly technological society. Students are expected to be literate in the STEM fields to successfully compete in the global economy (Basham, et al., 2010).

Properly preparing students will increase the workforce. The United States will be better prepared to compete in a globalized market with a STEM literate workforce. This preparation should increase students from the United States entering and graduating from institutions of higher education in STEM fields helping to alleviate STEM shortages. Another solution for alleviating potential shortages is to increase the number of students from underrepresented populations pursuing degrees in STEM (Lam, et al., 2008; National Science Foundation, 2000).

The Need for STEM Education in Special Education Curriculum and Instruction

Barriers to STEM Participation

Several groups are underrepresented in the STEM areas (Committee on Equal Opportunities in Science and Engineering, 2006; Dunn, et al., 2012), including students with disabilities (Alston & Hampton, 2000; Burgstahler & Doyle, 2005; National Science Foundation, 2000). The under-representation of people with disabilities in STEM careers is not surprising. Dunn and colleagues (2012) document several barriers to STEM careers for people with disabilities: (a) limited educational experiences, (b) preparation for STEM careers, and (c) employment issues. There is a lack of STEM role models for students with disabilities (Alston & Hampton, 2000). In addition, parent and teacher misconceptions that students with disabilities cannot be successful in STEM are a hindrance resulting in a lack of encouragement to take courses in these areas. Students with disabilities are not provided appropriate information or counseling related to taking advanced courses or STEM career preparation. Along these same lines teachers' lack of knowledge and skills regarding how to include students with disabilities serves as a barrier (Alston & Hampton, 2000). Other factors include technical barriers to science education and lower participation rates in structured and unstructured STEM-related activities (Dunn, et al., 2012).

Similarly, Burgstahler and Doyle (2005) endorsed many reasons for why people with disabilities do not pursue STEM careers citing: (a) low expectations and lack of encouragement from influential adults including teachers; (b) the lack of knowledge of educators about the content; (c) requirements of STEM content; and (d) careers or the technology available to increase access to STEM content to pursue STEM careers (Alston & Hampton, 2000). Echoing these sentiments Alston, Hampton, Bell, and Strauss (1998) found counselors perceived students with disabilities lacked the education skill required for careers in STEM, recommended careers other than STEM for students with disabilities, and believed that parents and teachers of students with disabilities suggested courses and careers other than STEM (Alston & Hampton, 2000). Furthermore, parents view indifference by educational professionals and overall school cultures as reasons for students with disabilities steer away from STEM content and careers (Alston & Hampton, 2000).With these barriers students with disabilities may not view STEM content or careers as obtainable goals and therefore are not prepared for postsecondary opportunities. Other barriers include inaccessible facilities, curriculum materials, computers, equipment, or other resources (Burgstahler & Doyle, 2005).

The National Science Foundation (NSF) supports these assertions citing lack of role models, academic tracking, and stereotyping have limited opportunities for people with disabilities in STEM. In addition, the degree of encouragement from parents and teachers, and courses taken in high school similarly influence whether people with disabilities select STEM careers (Alston & Hampton, 2000; National Science Foundation, 2000). Mastropieri and Scruggs (1992) found students with disabilities are not provided access to science curricula that equaled the curricula their non-disabled counterparts were provided (Alston & Hampton, 2000).

Adding to these barriers transitions are not always planned well for students with disabilities seeking to participate in postsecondary preparation (Burgstahler & Doyle, 2005). Support systems available in high school cease after graduation, and many students with disabilities lack the self-determination, college and employment preparation and independent living skills necessary to make successful transitions to adulthood. Students with disabilities continue to live with their parents or in other dependent living situations after high school more often than their non-disabled counterparts. Students with disabilities also experience isolation as they typically do not engage in social activities at the same rate as their non-disabled peers (Blackorby & Wagner, 1996; Harris & Associates, 1990).This social isolation has a negative effect on personal, academic, and career success (Burgstahler & Doyle, 2005;

Seymour & Hunter, 1998; Smith & Nelson, 1993) hampering self-determination and career aspirations.

Despite these barriers the number of students with disabilities participating in postsecondary education is increasing (Burgstahler & Chang, 2009; Burgstahler & Doyle, 2005; Henderson, 1999, 2001). Even with these increases students with disabilities face more challenges at the postsecondary and career level. Postsecondary experiences report that individuals with disabilities experience far less career achievement than their non-disabled counterparts. Adults with disabilities are more likely to be unemployed and experience underemployment, and work in low socioeconomic status jobs. (Burgstahler & Doyle, 2005). Individuals with disabilities are underrepresented in STEM professions, and scientists and engineers with disabilities experience higher unemployment rates than do other scientists and engineers (Burgstahler & Doyle, 2005; National Science Foundation, 2000).

The reasons for underrepresentation of people with disabilities in STEM careers are understandable in light of the challenges presented. These challenges begin in P12 settings. Students with disabilities often struggle with STEM content. Often students lack a solid foundation in STEM content which is evident upon entering secondary settings when intensity of instruction increases and weak foundations evidenced by misconceptions about scientific phenomena and misunderstanding of inquiry (Jacobson & Archodidou, 2000; Marino & Beecher, 2010; Mastropieri et al., 2006).

As students progress through school and struggle with weak foundations in critical content their attitudes are affected. Students develop negative attitudes toward STEM subjects as they encounter increasingly complex expository texts and other instructional materials that reduce their ability to access and comprehend scientific information (Basham, et al., 2010; M. Lee & Erdogan, 2007; Marino & Beecher, 2010). These experiences can make individuals with disabilities feel isolated coupled with a lack of potential, mentors, and peers who face challenges at school or work that are similar to their own (Alston & Hampton, 2000; Burgstahler & Doyle, 2005). In addition, Dalton, Moroco, Tivnan, and Mead (1997) found students with disabilities (Marino & Beecher, 2010):

- struggle with activating prior knowledge,
- are reluctant to pose questions,
- are less likely to have a systematic plan to approach problems,
- struggle to take instructor feedback,
- have difficulty making inferences during reasoning processes,
- seldom transfer knowledge across contexts
- are less likely than peers to be aware of their metacognitive processes

Samsonov, Pederson, and Hill (2006) added that students with disabilities require support to manage the extensive information imposed in advanced classes in addition to the supplementary secondary school features of rotating class schedules, limited instructional diversity, and teachers with inadequate knowledge of effective practices for teaching students with disabilities (Alston & Hampton, 2000; Alston, et al., 1998; Marino & Beecher, 2010; Mastropieri, et al., 2006; Robinson, 2002). STEM classrooms often fail to provide the level of instructional support that is necessary to promote successful learning outcomes for students with disabilities (Basham, et al., 2010; Langley-Turnbaugh, Wilson, & Lovewell, 2009).

Commonly utilized text-based instruction at the secondary level provides little support to students who struggle with learning due to limited reading skills or deficits in background knowledge (Basham, et al., 2010; Crockett, 2004; Scruggs, Mastropieri, & Okolo, 2008) adding to the struggles of students with disabilities in STEM classrooms. Students with different disabilities may have different experiences. Teachers can see the challenges that students with physical or sensory disabilities experience; however teachers have difficulty seeing the challenges faced by students experiencing challenges with cognitive processes. Students with hidden disabilities are at a disadvantage as people may not view these disabilities as "real" and perceiving these individuals as lazy, and told they are trying to use their disability to get out of work, or just seen as incapable (Denhart, 2008; A. Lee, 2011; Wolanin & Steele, 2004). As a result of misperceptions students with hidden disabilities are not encouraged to consider or prepare for careers in the STEM (Dunn, et al., 2012).

Increasing Access to STEM Content

Reasons for the struggles of students with disabilities in STEM content can be identified and addressed. Alston & Hampton (2000) found teachers' acknowledged a lack of understanding of the needs of people with disabilities which demonstrates a need to educate professionals on how to effectively incorporate students with disabilities into science and mathematics classes. Kimmel, Deek, O'Shea, and Farrell (1999) found teachers demonstrated knowledge to discuss generic adaptations and modifications to specific situations for students, but had difficulty in applying them to specific situations for different students. Kimmel and colleagues (1999) stated teacher efficacy can be a big factor in shaping teacher willingness and ability to make appropriate adaptations in instruction. Teachers were concerned about modifying practices when working with large groups of students. Marino and Beecher (2010) emphasized teachers must understand that complex concepts requires time, scaffolds, and engaging experiences. In addition to these classroom related concerns, Alston and Hampton (2000) reported parents and teachers believe students with disabilities are being directed away from STEM careers. With the challenging experiences and inadequate preparation, it is not surprising that students with disabilities are not prepared for a STEM workforce (Burgstahler & Chang, 2009). When providing instruction for students with disabilities in STEM content teachers should remember many students learn in nontraditional ways and an assortment of teaching methods should be utilized (Dunn, et al., 2012; Sternberg & Grigorenko, 2002). Many teachers have difficulty adapting instruction and materials to assist students with retention and application of knowledge. Teachers also struggle with providing appropriate accommodations for students with disabilities in STEM content. Instituting the use of graphic organizers, appropriate technology, and universal design are examples of how to increase access to content for students with disabilities (Dunn, et al., 2012; Rule, Stefanich, Hadelhuhn, & Peiffer, 2009).

Universal Design for Learning

Universal design for learning (UDL), technology, and accommodations have been found to be beneficial for participation in STEM curricula by students with disabilities (AccessSTEM, 2007; Dunn, et al., 2012). UDL for learning encourages teachers to become familiar with the recognition networks by providing multiple means of representation (presenting information and content in different ways), strategic networks by providing multiple means of action and expression (differentiating the ways that students can express what they know), and affective networks of learning by providing multiple means of engagement (stimulating interest and motivation for learning) (National Center on Universal Design for Learning, 2013). UDL principles have found to benefit all students, including students with disabilities. Even though UDL can increase access to STEM content for students with disabilities the use of UDL does not eliminate the need for specific accommodations for students with disabilities (Burgstahler, 2012).

Technology

Technology is also found to be an effective practice to increase the participation of students with disabilities in STEM content (AccessSTEM, 2007; Dunn, et al., 2012). Computers, assistive technology (AT), and network resources can link the communication and accessibility for students with disabilities (Burgstahler, 1994; Dunn, et al., 2012). In addition, programs such as text-to-speech, speech-to-text, and even spell check can assist students in being successful in their academic pursuits. Marino and Beecher (2010) encouraged the use of technology, specifically gaming, in combination with UDL. The combination serves as a compensatory scaffold to support students with disabilities in accessing science content (Bull & Bell, 2008; Marino & Beecher, 2010). Furthermore, the use of gaming technology can provide real time data for instructional support and remedial instruction within the game (Marino & Beecher, 2010). Marino and Beecher (2010) supported the use of gaming as a technology-based scaffold that can engage all students and meet them at their ability level. Additionally, Burgstahler and Doyle (2005) discussed the use of computer-mediated communication (CMC) to increase access to STEM. CMC is a process in which people use computers and networking technologies to communicate with one another, it may assist with communication with mentors or peers across space and time.

A subcategory of technology, assistive technology, can be a beneficial support for students with disabilities. Marino and Beecher (2010) encouraged the use of compensatory assistive technologies. Providing students with compensatory assistive technologies can improve the accessibility of leaning materials, promote STEM literacy, and improve students' attitudes and self-determination toward participating in STEM content and careers (AccessSTEM, 2007). The use of compensatory technologies provides students with disabilities access to sophisticated STEM content vocabulary and may include items such as electronic books and text software (Marino & Beecher, 2010).

HANDS-ON INSTRUCTION

Another strategy that can increase access to STEM for students with disabilities is hands-on instruction. Hands-on instruction and applied experiences are acknowledge as effective strategies for STEM instruction (Burgstahler & Chang, 2009; Dunn, et al., 2012; Liston, Peterson, & Ragan, 2007; Malian, 2007). Hands-on instruction can be formal or informal taking place in structured environments such as the classroom or more nontraditional environments such as extra-curricular activities, camps, or after school programs. These experiences provide opportunities for students to engage in extensions of academic activities, explore career paths, develop confidence in their abilities, and develop a network of other students with similar interest and goals as well as earn recognition for skills through awards and scholarships (Dunn, et al., 2012). Not only do they increase content knowledge and leaning, but they also contribute to increased self-confidence and career knowledge, motivation, and access (Dunn, et al., 2012; Lam, et al., 2008; Malian, 2007; Mastropieri & Scruggs, 1992).

Additional Strategies

In addition to these strategies there are some other general strategies that can increase the access to STEM curriculum for students with disabilities. Strategies such as explicit vocabulary instruction (Beck & McKeown, 2007; Scruggs & Mastropieri, 2000), anchored instruction (Glaser, Rieth, Kinzer, Colburn, & Peter, 2000) text adaptations (Bergerud, Lovitt, & Horton, 1988), and other content area reading strategies have been shown to improve student outcomes related to content specific STEM learning (Basham, et al., 2010). Marino and Beecher (2010) encouraged teachers to help students develop organizational clarity and to use exemplar models that are combined with explicit instruction. Mastropieri and Scruggs (2004) supported the use of differentiating instructional materials and the use of peer partners (Marino & Beecher, 2010). Brozo (2010) advocated connecting content to students' interest and outside-school competencies (Marino & Beecher, 2010). Marino and Beecher (2010) highlighted gaming as a way to engage students with disabilities in STEM content. The National Science Foundation supports the use of peer-peer and mentor-protégé networks for support and informal instruction are promising practices for increasing the participation of underrepresented groups in STEM fields (Burgstahler & Chang, 2009). Burgstahler and Doyle (2005) emphasized peer and mentor support; computer-mediated communication; hands-on science experiences in pre-college environments; work-based and research experiences; bridge programs between academic levels; tutoring; and preparation of educators and support staff.

Individualized Considerations

As emphasized in special education and individualized education programs, individually designed learning experiences should be provided. These experiences should include academic content, but also experiences outside the classroom, including extracurricular activities, after-school programs, mentors, special programs, and camps (Davis & Hardin, 2013; Dunn, et al., 2012); as well as reflected in IEP goals (Lam, et al., 2008). Dunn (2012) stated these activities should promote students':

- interest and motivation in STEM;
- deep understanding of mathematics, science and other content;
- ways to apply knowledge and skill creatively in problem-solving situations; and
- understanding that persons with disabilities can indeed be successful in STEM careers.

Early Exposure

Research suggests that these experiences must begin early (Burgstahler & Chang, 2009; Jacobs & Eccles, 1992; Lam, et al., 2008; Simpkins, Davis-Kean, & Eccles, 2006). Moomaw and Davis (2010) suggested implementing STEM activities in the early childhood classroom indicating that appropriate STEM activities permit young children to use all their senses to explore materials and helps children focus, increase their vocabulary, collaborate with one another and create scientific relationships, in addition to young children being ready, willing and able to engage in STEM activities. Dunn (2012) emphasized this need to start early stating no later than middle school because early learning experiences play a critical role in career development (Dunn, et al., 2012; Lam, et al., 2008; Lent, Brown, & Hackett, 1994, 2000; Malian, 2007) Early experiences "shape self-efficacy, beliefs, and outcome expectations, which in turn affect the formation of vocational interest, which subsequently influence occupational goals, choice actions and performance attainments." (Alston & Hampton, 2000; Dick & Rallis, 1991; Dunn, et al., 2012; Lam, et al., 2008; Lent, et al., 1994, 2000; Malian, 2007). In short Dunn and colleagues (2012) suggested that a foundation for postsecondary education in STEM be made early and include personal development, access to content, experiential development, and postsecondary connecting activities.

Transition Considerations

Students must be prepared to and supported through transitions from secondary to postsecondary/adulthood (Burgstahler & Chang, 2009; Burgstahler & Doyle, 2005; Kohler & Chapman, 1999; National Council on Disability and Social Security Administration, 2000). Students transition planning must also reflect their interest STEM. Dunn (2012) suggested building student confidence and confidence in STEM by laying the foundation for postsecondary education in STEM in their transition consideration and ensuring students':

- Personal development,
- Access to content,
- Experiential development; and
- Postsecondary connecting activities.

Furthermore importance lies in students being able to communicate their needs and identify appropriate accommodations in order to persist in STEM programs (Dunn, et al., 2012).

BENEFITS OF STEM EDUCATION IN SPECIAL EDUCATION CURRICULUM AND INSTRUCTION

People with disabilities are underrepresented in STEM careers, but the presence of people with disabilities in STEM education and careers is not a new phenomenon. There are documented benefits of people with disabilities participation in STEM curriculum and careers (Kirch, Bargerhuff, Cowan, & Wheatly, 2007; Marino & Beecher, 2010). Moomaw and Davis (2010) found benefits of including STEM in early childhood curriculum. When exposing preschool children to STEM activities the children constructed important scientific and mathematical relationships. They explored materials that piqued their scientific curiosity and math discovery. They suggest that these children can build upon these foundational experiences. These experiences can teach young children that math and science can be exciting.

Cawley, Hayden, Cade, and Baker-Kroczynski (2002) found that students with disabilities passed district science exams at the same rate as peers without disabilities after teachers were provided with extensive preparation focusing on effective instruction (Basham, et al., 2010; Marino & Beecher, 2010). Kirch, Bargerhuff, Cowan, and Wheatly (2007) found that when general education science teachers were able to eliminate curriculum barriers, they were overwhelmed by students' abilities to produce meaningful learning outcomes in inclusive settings (Marino & Beecher, 2010).

Case Study

One great example of a program that is preparing people with disabilities to enter STEM careers is the DO-IT Scholars program which is out of the DO-IT Center at the University of Washington in Seattle, Washington (DO-IT Center, 2006). DO-IT stands for Disabilities, Opportunities, Internetworking, and Technology. The DO IT Scholars Program serves students with a range of disabilities who are college-bound and interested in STEM careers. The program includes well-defined and established components. The program was established in 1993 and has sustained programming for over 20 years (DO-IT Center: University of Washington, 2013).

The DO-IT Center's Scholars program employs three interventions with participants. Students participate in a summer study which is a residential summer study session at the University of Washington where participants are engaged in college, career, and leadership training. This assists with transition issues that may arise for students as they transition from high school to postsecondary and from postsecondary to career (Burgstahler & Chang, 2009; DO-IT Center: University of Washington, 2013). In addition to the summer study, students engage in year-round computer internet activities. These activities assist in participants' development of computer and internet skills to support academic and career aspirations.

Communication between mentors and peers is also encouraged through the year-round computer and internet activities (Burgstahler & Chang, 2009). The final component of the Scholars program involves work experiences. Participants are engaged in internships and other work-based learning activities. Through these experiences students are encouraged to explore their interests and develop their skills. In addition students are encouraged to self-advocacy by disclosing their disabilities and seeking accommodations, using assistive technology at work sites, and working with supervisors and colleagues (Burgstahler & Chang, 2009). In short the program seeks to develop participants' self-determination and self-advocacy abilities while developing their academic, social, and career skills (Burgstahler & Chang, 2009). Mentors are an important component of the program. Mentors are adults who provide young people with disabilities with advice, information, contacts, support, and role models (DO-IT Center: University of Washington, 2013).

There are numerous documented benefits of this program. Parents of participants found their children's interest in college increased. In addition to this, parents reported an increase in their child's awareness of college options, self-esteem, self-advocacy as well as social, academic and career skills (Burgstahler, 2002; Burgstahler & Chang, 2009). Scholars of the DO-IT program found the program beneficial and reported similar sentiments of their parents. These benefits included feeling more prepared for college and employment, development of internet, computer, social, self-advocacy, self-esteem, perseverance and independent living skills. Students also felt more aware of career options (Burgstahler, 2003; Burgstahler & Chang, 2009; Kim-Rupnow & Burgstahler, 2004).

Participants found the most beneficial aspect of the summer study program was the development of social skills, followed by the development of academic and career skills (Burgstahler, 2003; Burgstahler & Chang, 2009; Kim-Rupnow & Burgstahler, 2004). Participants reported the most beneficial effect of the year-round computer and internet activities was the development of career skills followed by academic and social skills (Burgstahler, 2003; Burgstahler & Chang, 2009; Kim-Rupnow & Burgstahler, 2004). In addition, scholars found email very helpful for staying close to friends and family and meeting people from around the world. Email permitted a quick, easy and inexpensive method of communication with many at once. Email also assisted with getting answers to specific questions. Another benefit espoused by participants was the ability to communicate independently without disclosing their disabilities (Burgstahler & Chang, 2009; Burgstahler & Cronheim, 2001; Burgstahler & Doyle, 2005).

Participants found the work-based learning opportunities increased their motivation to pursue STEM careers. The experiences increased their knowledge about careers and the workplace, increased their job-related skills and ability to work with supervisors and coworkers. Students felt empowered to self-advocate for accommodations and other needs (Burgstahler, 2001; Burgstahler, Bellman, & Lopez, 2004; Burgstahler & Chang, 2009).

DO-IT Scholars predicted that the access to internet and computer activities provided by the programs would contribute to their success in college and future careers and STEM. Scholars also found peer and mentor relationships beneficial by providing psychosocial, academic, and career support which furthered their academic and career interests (Burgstahler, 2003; Burgstahler & Chang, 2009; Burgstahler & Cronheim, 2001; Burgstahler & Doyle, 2005; Kim-Rupnow & Burgstahler, 2004). For the mentors in particular, participants indicated mentors stimulated interest in STEM (Burgstahler & Chang, 2009; Burgstahler & Cronheim, 2001; Burgstahler & Doyle, 2005).

Mentors shared topics discussed with their scholars. Topics discussed reflect benefits espoused by participants, including STEM, college issues, disability-related issues, and careers (Burgstahler & Chang, 2009; Burgstahler & Cronheim, 2001). Burgstahler and Chang (2009) encouraged the programs to model the DO-IT Scholars program as the program has found success increasing STEM participation by people with disabilities. The authors inspired programs to include those students who are not initially interested in STEM, as these students have increased their interest in STEM through exposure to program components. Importance lies in providing interventions tailored to the interests of those students and increasing opportunities to augment awareness of the wide variety of STEM careers, interest in STEM, and confidence in pursuing STEM fields.

CONCLUSION

Students with disabilities are underrepresented in STEM careers and are not typically encouraged to pursue STEM careers or to enroll in STEM courses. Career achievements of people with disabilities suggest the potential of this population to be successful in STEM careers and diversify the STEM field (Blumenkopf, Stern, Swanson, & Wohlers, 1996; Burgstahler, et al., 2004; Burgstahler & Doyle, 2005; Unger, Wehman, Yasuda, Campbell, & Green, 2001). The potential of students with disabilities to excel in STEM careers has not been fully realized or encouraged. With the advances in assistive technology STEM careers are becoming more accessible to people with disabilities (Burgstahler & Doyle, 2005). With proper accommodations and improved attitudes and perceptions in the education system and workplace could help change the current climate (Alston & Hampton, 2000). This climate change can begin by including students with a range of disabilities in STEM education.

REFERENCES

AccessSTEM. (2007). Building capacity to include students with disabilities in science, technology, engineering, and mathematics fields. Seattle: University of Washington.

Alston, R. J., & Hampton, J. (2000). Science and engineering as viable career choices for students with disabilities: A survey of parents and teachers. *Rehabilitation Counseling Bulletin*, 43(3), 158-164.

Alston, R. J., Hampton, J. L., Bell, T. J., & Strauss, M. (1998). Matriculation of persons with disabilities in science and engineering: Perceptions of rehabilitation counselors. *Journal of Applied Rehabilitation Counseling*, 29(3), 5-8.

Basham, J. D., Israel, M., & Maynard, K. (2010). An ecological model of STEM education: Operationalizing STEM FOR ALL. *Journal of Special Education Technology*, 25(3), 9-19.

Beck, I. L., & McKeown, M. G. (2007). Increasing young chldren's oral vocabulary repertories through rich and focused instruction. *Elementary School Journal*, 107, 251-271.

Bergerud, D., Lovitt, T. C., & Horton, S. (1988). The effectiveness of textbook adaptations in life science for high school students with Learning Disabilities. *Journal of Learning Disabilities,* 21(2), 70-76.

Blackorby, J., & Wagner, M. (1996). Longitudinal post-school outcomes of youth with disabilities: Findings form the National Longitudinal Transition Study. *Exceptional Children,* 62(5), 399-413.

Blumenkopf, T., Stern, V., Swanson, A., & Wohlers, D. (Eds.). (1996). Working chemists with disabilities: Expanding opportunities in science: American Chemical Society.

Brozo, W. G. (2010). Response to intervention or responsive instruction? Challenges and possibilities of response to intervention for adolescent literacy. *Journal of Adolescent and Adult Literacy*, 53(4), 277-281.

Bull, G., & Bell, R. L. (2008). Education technology in the science classroom. In R. L. Bell, J. Gess-Newsome & J. Luft (Eds.), Technology in the secondary science classroom. Arlington, VA: NSTA Press.

Burgstahler, S. (1994). Increasing the representation of people with disabilities in science, engineering, and mathematics. *Journal of Information Technology for Development*, 4(9), 1-8.

Burgstahler, S. (2001). A collaborative model promotes career success for students with disabilities: How DO-IT does it. *Journal of Vocational Rehabilitation*, 16(129), 1-7.

Burgstahler, S. (2002). The value of DO-IT to kids who did it! Exceptional Parent, 32(11), 79-86.

Burgstahler, S. (2003). DO-IT: Helping students with disabilities transition to college and careers. *National Center on Secondary Education and Transition Research to Practice Brief*, 2(3).

Burgstahler, S. (2012). Equal access: Universal design of instruction. Retrieved from http://www.washington.edu/doit/Brochures/Academics/equal_access_udi.html

Burgstahler, S., Bellman, S., & Lopez, S. (2004). Research to practice: DO-IT prepares students with disabilities for employment. *NACE Journal*, 65(1).

Burgstahler, S., & Chang, C. (2009). Promising interventions for promoting STEM fields to students who have disabilities. *Review of Disability Studies An International Journal*, 5(2), 29-47.

Burgstahler, S., & Cronheim, D. (2001). Supporting peer-peer and mentor-protege relationships on the internet. *Journal of Research on Technology in Education*, 34(1), 59-74.

Burgstahler, S., & Doyle, A. (2005). Gender differences in computer-mediated communication among adolescents with disabilities: A case study. Disability Studies Quarterly, 25(2). Retrieved from http://dsq-sds.org/article/view/552/729

Cawley, J., Hayden, S., Cade, E., & Baker-Kroczynski, S. (2002). Including students with disabilities into the general education scienc classroom. *Exceptional Children*, 68, 423-435.

Committee on Equal Opportunities in Science and Engineering. (2006). 2005-2006 Biennial report to Congress. Arlington, VA: National Science Foundation.

Committee on Science and Public Policy. (2005). Rising above the gathering storm: Energizing and employing America for a brighter economic future. Washington, DC: National Academies Press.

Crockett, J. B. (2004). Taking stock of science in the schoolhouse: Four ideas to foster effective instruction. *Journal of Learning Disabilities*, 37(3), 189-199.

Dalton, B., Morocco, C., Tivnan, T., & Mead, P. (1997). Supported inquiry science: Teaching for conceptual change in urban and suburban classrooms. *Journal of Learning Disabilities,* 30, 670-684.

Davis, K. E. B., & Hardin, S. E. (2013). Making STEM fun: How to organize a STEM camp. *Teaching Exceptional Children*, 45(4), 60-67.

Denhart, H. (2008). Deconstructing barriers: Perceptions of students labeled with Learning Disabilities in higher education. *Journal of Learning Disabilities*, 41(6), 483-497.

Dick, T. P., & Rallis, S. F. (1991). Factors and influences on high school students' career choices. Journal for Research in Mathematics Education, 22, 281-292.

DO-IT Center. (2006). DO-IT Scholars Retrieved May 13, 2013, from http://www.washington.edu/doit/Brochures/PDF/scholars.pdf

DO-IT Center: University of Washington. (2013). DO-IT Retrieved May 13, 2013, from http://www.washington.edu/doit/

Dunn, C., Rabren, K. S., Taylor, S. L., & Dotson, C. K. (2012). Assisting students with high-incidence disabilities to pursue careers in science, technology, engineering, and mathematics. *Intervention in School and Clinic*, 48(1), 47-54.

Fleischman, H. L., Hopstock, P. J., Pelczar, M. P., Shelley, B. E., & Xie, H. (2010). Highlights from PISA 2009: Performance of U.S. 15-year-old students in reading, mathematics, and science literacy in an international context (NCES 2011-004). U.S. Department of Education, National Center for Education Statistics. Washington, DC: Government Printing Office.

Friedman, T. L. (2006). The world is flat: A brief history of the twenty-first century. New York: Farrar, Straus and Giroux.

Gilman, D. (1898). University problems in the United States. New York: Century Company.

Glaser, C. W., Rieth, H. J., Kinzer, C. K., Colburn, L. K., & Peter, J. (2000). A description of the impact of multimedia anchored instruction on classroom interactions. *Journal of Special Education Technology*, 14(2), 27-43.

Harris, L., & Associates. (1990). National organization on disability/Harris survey of Americans with disabilities. New York: Author.

Henderson, C. (1999). College freshman with disabilities: Statistical year 1998. Washington, DC: American Council on Education.

Henderson, C. (2001). College freshman with disabilities: A biennial statisical profile. Washington, DC: American Council on Education.

Jacobs, J. E., & Eccles, J. S. (1992). The impact of mothers' gender-role sterotypic beliefs on mothers' and children's ability perceptions. *Journal of Personality and Social Psychology*, 63, 932-944.

Jacobson, M. J., & Archodidou, A. (2000). The design of hypermedia tools for learning: Fostering conceptual chane and transfer for complex knowledge. *The Journal of the Learning Sciences*, 9, 145-199.

Kim-Rupnow, W. S., & Burgstahler, S. (2004). Perceptions of students with disabilities regarding the valude of technology-based support activities on postsecondary education and employment. *Journal of Special Education Technology*, 19(2), 43-56.

Kimmel, H., Deek, F. P., Farrell, M. L., & O'Shea, M. (1999). Meeting the needs of diverse student populations: Comprehensive professional development in scienc, math, and technology for teachers of students with disabilities. *School Science and Mathematics*, 99(5), 241-249.

Kirch, S. A., Bargerhuff, M. E., Cowan, H., & Wheatly, M. (2007). Reflections of educators in pursuit of inclusive science classrooms. *Journal of Science Teacher Education*, 18, 663-692.

Kohler, P. D., & Chapman, S. (1999). Literature review on school-to-work transition. Champaign, IL: University of Illinois at Urbana-Champaign, Transition Research Institute.

Lam, P., Doverspike, D., Zhao, J., Zhe, J., & Menzemer, C. (2008). An evaluation of a STEM program for middle school students on Learning Disability related IEPs. *Journal of STEM Education*, 9(1&2), 21-29.

Langley-Turnbaugh, S. J., Wilson, G., & Lovewell, L. (2009). Increasing the accessibility of science for all students. *Journal of Science Education for Students with Disabilities*, 13, 1-8.

Lantz, H. B. (2009). Science, Technology, Engineering, and Mathematics (STEM) Education What Form? What Function? Retrieved from http://www.currtechintegrations.com/pdf/STEMEducationArticle.pdf

Lee, A. (2011). Postsecondary science, engineering, and mathematics (STEM) enrollment comparisons for students with and without disabilities. *Career Development for Exceptional Individuals*, 34(2), 72-82.

Lee, M., & Erdogan, I. (2007). The effect of science-technology-society teaching on students' attitudes toward science and certain aspects of creativeity. *International Journal of Science Education*, 11, 1315-1327.

Lent, R. W., Brown, S. D., & Hackett, G. (1994). Toward a unifying social cognitive theory of career and academic interest, choice, and performance. *Journal of Vocational Behavior*, 45, 79-122.

Lent, R. W., Brown, S. D., & Hackett, G. (2000). Contextual supports and barriers to career choice: A social congitive analysis. *Journal of Counseling Psychology*, 47, 36-49.

Liston, C., Peterson, K., & Ragan, V. (2007). Guide to promising practices in information technology education for girls. Boulder, CO: National Center for Women and Information Technology.

Malian, I. (2007). MVIP: Math villages for inclusive practices: A model to engage all students and teachers in STEM experiences. Electronic Journal for Inclusive Education, 7(2). Retrieved from http://www.cehs.wright.edu/resources/publications/ejie/WinterSpring2011/JIE%20MVIP.htm

Marino, M. T., & Beecher, C. C. (2010). Conceptualizing RTI in 21st-century secondary science classrooms: Video games' potential to provide tiered support and progress monitoring for students with Learning Disabilities. *Learning Disability Quarterly*, 33(4), 299-311.

Mastropieri, M. A., & Scruggs, T. E. (1992). Science for students with disabilities. *Review of Educational Research*, 62, 377-411.

Mastropieri, M. A., & Scruggs, T. E. (2004). Science and schooling for students with LD: A discussion of the symposium. Journal of Learning Disabilities, 37, 270-276.

Mastropieri, M. A., Scruggs, T. E., Norland, J. J., Berkeely, S., McDuffie, K., Tornquist, E. H., & Connors, N. (2006). Differentiated curriculum enhancement in inlcusive middle school science: Effects on classroom and high-stakes tests. *The Journal of Special Education,* 40, 130-137.

Moomaw, S., & Davis, J. A. (2010). STEM comes to preschool. *Young Children*, 65(5), 12-14, 16-18.

Morrison, J. S. (2006). TIES STEM Education Monograph Series. Attributes of STEM education. Baltimore, MD: TIES.

National Center on Universal Design for Learning. (2013). About UDL Retrieved May 13, 2013, from http://www.udlcenter.org/aboutudl/whatisudl

National Council on Disability and Social Security Administration. (2000). Transition and post-school outcomes for youth with disabilities: Closing the gaps to post-secondary education and emplyment. Washington, DC: Author.

National Governors Association & Council of Chief State School Officers. (2013). Common Core State Standards Initiative, Retreived from http://www.corestandards.org/ Math/Practice

National Science Board. (2006). Scince and engineering indicators 2006 (Vol. 1, NSB06-01; Vol. 2, NSB 06-01A). Arlington, VA: Author.

National Science Foundation. (2000). Women, minorities, and persons with disabilities in science and engineering. Washington, DC: U.S. Government Printing Office.

National Science Teachers Association. (2013). Retrieved from http://www.nsta.org/

Office of Disability Employment Policy. (2001). Improving the availability of community-based services for people with disabilities. Washington, DC: Author.

Provasnik, S., Kastberg, D., Ferraro, D., Lemanski, N., Roey, S., & Jenkins, F. (2012). Highlights from TIMSS 2011 mathematics and science achievement of U.S. fourth- and eighth-grade students in an international context (NCES 2013-009): National Center for Education Statistics, Institute of Education Sciences, U.S. Department of Education. Washington, DC.

Robinson, S. (2002). Teaching high school students with learning and emotional disabilities in inclusion science classrooms: A case study of four teachers' beliefs and practices. *Journal of Science Teacher Education*, 13, 13-26.

Rule, A., Stefanich, G., Hadelhuhn, C., & Peiffer, B. (2009). A working conference on students with disabilities in STEM coursework and careers. from ERIC database (ED505568).

Samsonov, P., Pedersen, S., & Hill, C. (2006). Using problem-based learning software with at-risk students: A case study. *Computers in the Schools*, 23(1/2), 111-124.

Scruggs, T. E., & Mastropieri, M. A. (2000). The effectiveness of mnemonic instruction for students with learning and behavior problems: An update and research synthesis. *Journal of Behavioral Education*, 10(2), 163-173.

Scruggs, T. E., Mastropieri, M. A., & Okolo, C. (2008). Science and social studies for students with disabilities. *Focus on Exceptional Children*, 41, 1-25.

Seymour, E., & Hunter, A. (1998). Talking about disability: The education and work experience of graduates and undergraduates with disabilities in science, mathematics and engineering majors (AAAS Publication No. 98-02S). Washington, DC: American Association for the Advancement of Science.

Simpkins, S. D., Davis-Kean, P. E., & Eccles, J. S. (2006). Math and science motivation: A longitudinal examination of the links between choices and beliefs. *Developmental Psychology*, 42(1), 70-83.

Smith, D. J., & Nelson, J. R. (1993). Factors that influence the academic success of college students with disabilities. Paper presented at the 71st Annual Convention of the Council for Exceptional Children, San Antonio, TX.

Sternberg, R. J., & Grigorenko, E. L. (2002). Difference scores in the identification of children with Learning Disabilities: It's time to use a different method. *Journal of School Psychology*, 40(1), 65-83.

Tsupros, N., Kohler, R., & Hallinen, J. (2009). STEM education: A project to identify the missing components. Pennsylvania: Leonard Gelfand Center for Service Learning and Outreach at Carnegie Mellon University and The Intermediate Unit 1 Center for STEM Education.

Tyson, W., Lee, R., Borman, K. M., & Hanson, M. A. (2007). Science, technology, engineering, and mathematics (STEM) pathways: High school science and math coursework and postsecondary degree attainment. *Journal of Education for Students Placed At Risk*, 12(3), 243-270.

Unger, D., Wehman, P., Yasuda, S., Campbell, L., & Green, H. (2001). Human resource professionals and the employment of persons with disabilities: A business perspective. Paper presented at the Capacity Building Institute, University of Hawaii.

Wolanin, T. R., & Steele, P. E. (2004). Higher education opportunities for students with disabilities: A primer for policy makers. Washington, DC: Institute for Higher Education Policy.

In: S.T.E.M. Education
Editor: Satasha L. Green

ISBN: 978-1-63117-806-1

Chapter 2

WORKING WITH LEARNERS WITH COMMUNICATION DISORDERS IN STEM

Sunday O. Obi*
Kentucky State University, US

ABSTRACT

Many students with communication disorders (CD) face challenges to acquiring speech or language. These delays and disorders may range from simple sound substitution to the inability to understand or use language. The intervention for children with CD must be specific to the individual child as well as promoting language development across the curriculum. For a student with CD to benefit from STEM, student and teacher collaboration must be an important component in any training the student receives in STEM.

Teachers must also make a sincere effort to learn from speech and language professionals and vice-versa in order for students with CD to achieve success. It is obvious that enhancing STEM programs to benefit students with CD requires a process of system change, as opposed to isolated programs and invalidated instructional practices often common with programming for these students. The overall purpose of this chapter is to present ideas that help STEM programs become more inclusive systems of education that strive to produce better outcomes for all students, especially students with communication disorders.

INTRODUCTION

Students with communication disorders (CD) have average to above average intelligence but have difficulties saying sounds correctly to being completely unable to speak or understand speech. Individuals with CD can be considered as that group of students in the education system who has an IEP that is based on a diagnosis consistent with an impairment in one's ability to talk, understand, read, and write. Students with CD can be provided

* E-mail: Sunday.obi@kysu.edu.

interventions by speech-language pathologist. For those that cannot speak at all, using speech generating devices will enhance communication.

All students, including students with significant CD, deserve and have a right to a quality educational experience. This right includes, to the maximum extent possible, the opportunity to be involved in and meet the same challenging expectations that have been established for all students. Because one has CD does not mean that his or her skills are not up to par. With differentiated instruction: extra accommodation, additional resources, and extra help after class and different tests or extra time on tests, students with CD will achieve success in STEM programs. Overall, the intent is to increase interest in STEM careers for students with and without Individual Education Plans (IEP).

Brief Summary of Communication Disorders and Prevalence

Children are eligible for services for a CD if they have "a communication disorder such as stuttering, impaired articulation, a language impairment or a voice impairment, which adversely affects a child's education performance". Communication disorders are impairments in the ability to receive, send, process, and comprehend concepts or verbal, nonverbal, and graphic symbol systems. A CD may be evident in the process of hearing, language, and/or speech (ASHA, 1993). Individuals may demonstrate one or any combination of communication disorders. A CD may result in a primary disability or it may be secondary to other disabilities.

Speech disorder: A speech disorder is an impairment of the articulation of speech sounds, fluency and/or voice. An articulation disorder is the atypical production of speech sounds characterized by substitutions, omissions, additions or distortions that may interfere with intelligibility.

Fluency disorder: A fluency disorder is an interruption in the flow of speaking characterized by atypical rate, rhythm, and repetitions in sounds, syllables, words, and phrases. This may be accompanied by excessive tension, struggle behavior, and secondary mannerisms.

Voice disorder: A voice disorder is characterized by the abnormal production and/or absences of vocal quality, pitch, loudness, resonance, and/or duration, which is inappropriate for an individual's age and/or sex. While a language disorder is impaired comprehension and/or use of spoken, written and/or other symbol systems. The disorder may involve (1) the form of language (phonology, morphology, and syntax), (2) the content of language (semantics), and/or (3) the function of language in communication (pragmatics) in any combination.

Form of Language: Phonology is the sound system of a language and the rules that govern the sound combinations. Morphology is the system that governs the structure of words and the construction of word forms. Syntax is the system governing the order and combination of words to form sentences, and the relationships among the elements within a sentence.

Content of Language: Semantics is the system that governs the meanings of words and sentences.

Function of Language: Pragmatics is the system that combines the above language components in functional and socially appropriate communication.

Hearing Disorder

A hearing disorder is the result of impaired auditory sensitivity of the physiological auditory system. A hearing disorder may limit the development, comprehension, production, and/or maintenance of speech and/or language. Hearing disorders are classified according to difficulties in detection, recognition, discrimination, comprehension and perception of auditory information. Individuals with hearing impairments may be described as deaf or hard of hearing.

Deaf: A hearing disorder that limits an individual's aural/oral communication performance to the extent that the primary sensory input for communication may be other than the auditory channel.

Hard of Hearing: A hearing disorder, whether fluctuating or permanent, which adversely affects an individual's ability to communicate. The hard-of-hearing individual relies on the auditory channel as the primary sensory input for communication.

Central auditory processing disorders (CAPD) are deficits in the information processing of audible signals not attributed to impaired peripheral hearing sensitivity or intellectual impairment. This information processing involves perceptual, cognitive, and linguistic functions that, with appropriate interaction, result in effective receptive communication of auditorily presented stimuli. Specifically, CAPD refers to limitations in the ongoing transmission, analysis, organization, transformation, elaboration, storage, retrieval, and use of information contained in audible signals. CAPD may involve the listener's active and passive (e.g., conscious and unconscious, mediated and unmediated, controlled and automatic) ability to do the following:

(1) attend, discriminate, and identify acoustic signals; (2) transform and continuously transmit information through both the peripheral and central nervous systems; (3) filter, sort, and combine information at appropriate perceptual and conceptual levels; (4) store and retrieve information efficiently; (5) restore, organize, and use retried information; (6) segment and decode acoustic stimuli using phonological, semantic, syntactic, and pragmatic knowledge; and (7) attach meaning to a stream of acoustic signals through use of linguistic and nonlinguistic contexts.

Communication Variations

Communication difference/dialect is a variation of a symbol system used by a group of individuals that reflects and is determined by shared regional, social, or cultural/ethnic factors. A regional, social, or cultural/ethnic variation of a symbol system should not be considered a disorder of speech or language. Augmentative/alternative communication systems attempt to compensate and facilitate, temporarily or permanently, for the impairment and disability patterns of individuals with severe expressive and/or language comprehension disorders. Augmentative/alternative communication may be required for individuals demonstrating impairments in general, spoken, and/or written modalities.

Estimates of the prevalence of CD in children vary widely. Reliable figures are hard to come by because investigators often employ different definitions of speech and language disorders and sample different populations. In the 2009-10 school year, 1,107,029 children ages 6-21 received special education services under the IDEA category of "speech or language impairments" (U.S. Department of Education, 2011). This number represents about 2.5% of the school-age population and 19% of all students receiving special education services, making speech or language impairments the second-largest category after learning disabilities.

The actual number of children with speech and language impairments is much higher. Approximately 50% of children who receive special education services because of another primary disability (e.g., intellectual disabilities, learning disabilities, hearing impairments) also have communication disorders (Hall et al., 2001).

School-based speech-language pathologists (SLPs) work with a median caseload of 50 students each month (ASHA, 2010). Approximately half of all elementary students who are served by SLPs have speech and language production problems. About 1 in 20 children served by SLPs has fluency disorders. Speech and language impairments are more prevalent among males than females and are about the same in each of the major geographic regions of the United States. Approximately two-thirds of school-age children served by SLPs are boys (Hall et al., 2001). The percentage of children with speech and language disorders decreases significantly from the earlier to the later school grades.

Many speech problems are developmental rather than physiological, and as such they respond to therapeutic instruction. Language experiences are central to a young child's development. In the past, children with CDs were routinely removed from the general education classroom for individual speech and language therapy. This is still the case in severe instances, but the trend is toward keeping the child in the mainstream as much as possible. In order to accomplish this goal, teamwork among the teacher, speech and language therapist, audiologist, and parents is essential. Speech improvement and correction are blended into the general education classroom curriculum and the child's natural environment.

Amplification may be extremely valuable for the child with a hearing impairment. Students whose hearing is not completely restored by hearing aids or other means of amplification have unique communication needs. Children who are deaf are not automatically exposed to the enormous amounts of language stimulation experienced by hearing children in their early years. For deaf children, early, consistent, and conscious use of visible communication modes such as sign language, finger spelling, and cued speech and/or amplification and aural/oral training can help reduce this language delay. Some educators advocate a strict oral approach in which the child is required to use as much speech as possible, while others favor the use of sign language and finger spelling, combined with speech, an approach known as total communication.

There is increasing consensus that whatever system works best for the individual should be used. Many children with hearing impairments can be served in the general education classroom with support services; instructional aides such as captioned films and high interest/low vocabulary reading materials. Students whose physical problems are so severe that they interfere with or completely inhibit communication can frequently take advantage of technological advances that allow the individual to make his or her needs and wants known, perhaps for the first time.

All students, including students with significant CDs, deserve and have a right to a quality educational experience. This right includes, to the maximum extent possible, the opportunity to be involved in and meet the same challenging expectations that have been established for all students. For teachers, students with CDs present a unique challenge. Many students with CDs have the ability to learn material close to (or even at) their grade level. However, these students may need a little extra support to achieve. In other words, STEM programs must redesign their curriculum in order to accommodate all students – general and special education students. That is to say, there are strategies and practices that STEM educators can do to increase the accessibility of STEM for students with communication disorders.

IMPORTANCE OF STEM INSTRUCTION FOR LEARNERS WITH COMMUNICATION DISORDERS

There is growing concern that the United States is not preparing a sufficient number of students, teachers, and professionals in the areas of science, technology, engineering, and mathematics (STEM). Although the most recent National Assessment of Educational Progress (NAEP) results show improvement in U.S. pupils' knowledge of math and science, the large majority of students still fail to reach adequate levels of proficiency. When compared to other nations, the achievement of U.S. pupils appears inconsistent with the nation's role as a world leader in scientific innovation. For example, among the 40 countries participating in the 2003 Program for International Student Assessment (PISA), the U.S. ranked 28th in math literacy and 24th in science literacy.

Some attribute poor student performance to an inadequate supply of qualified teachers. This appears to be the case with respect to subject-matter knowledge: many U.S. math and science teachers lack an undergraduate major or minor in those fields – as many as half of those teaching middle school math. Indeed, post-secondary degrees in math and physical science have steadily decreased in recent decades as a proportion of all STEM degrees awarded.

According to the National Science Foundation, the United States currently ranks 20th among all nations in the proportion of 24-year olds who earn degrees in national science or engineering. Once a leader in STEM education, the United States is now far behind many countries on several measures.

Advancements in technology and increased job specialization have resulted in career opportunities in fields that were once considered unattainable for individuals who have disabilities. Many of these careers require knowledge and skills obtained through postsecondary education. Although the number of individuals with disabilities seeking postsecondary education continues to increase, these students experience lower success rates than their non-disabled peers. Individuals with disabilities continue to be underrepresented in many challenging academic and career fields. Producing a sufficient pool of qualified graduates in the areas of STEM occupations has long been a challenge for American Universities and Colleges. The supply of the workforce in STEM areas is likely to get worse in the near future, as the number of students pursuing degrees appears to be shrinking. One solution to alleviating the potential shortages is to increase the number of students from underrepresented areas considering majoring in and pursuing careers in STEM (National Science Foundation, 2000).

One traditionally underrepresented group that has increased in size is students with CDs, Operationally, students with CDs can be considered as that group of individuals in the educational system who has an Individualized Education Plan (IEP) that is based on diagnosis consistent with an impairment in processes related to learning.

The Americans with Disabilities Act of 1990 and other federal and state legislation require that schools make programs accessible to students with disabilities. People with disabilities are underrepresented in many challenging careers, including STEM fields. Negative stereotyping and attitudes are reported as the most significant factors faced by people with disabilities in these fields. At the precollege level, students with disabilities are often discouraged from taking math and science courses and, when enrolled, are often not fully included in rigorous work in these classes. Poor high school preparation limits their options in college and careers.

One of the greatest difficulties facing people with disabilities is created when their capacity to be educated and function in the world is evaluated solely by limitations imposed by their disabilities. They essentially become defined by their disabilities, rather than their broad range of interests and abilities. For example, many people who sustain spinal cord injuries are encouraged to participate in wheelchair basketball after they leave rehabilitation facilities. This may be fine for some people, but for those who have never had an interest in playing basketball before, why should they become basketball players now? In an example like this, people are making assumptions about ability based solely upon the characteristics of a person's disability.

Another assumption is that a disability limits an individual in every aspect of his or her functioning, including the ability to be educated. To meet the intent of the Americans with Disabilities Act (ADA), educators must develop an attitude that is disability-positive. Students with disabilities can work hard and be productive and should be expected to do so. In order to reduce labor shortages in STEM occupations and to increase diversity in terms of backgrounds, it is necessary to target middle school children, with the intent of developing high school children with knowledge and confidence in STEM.

Social cognitive theories (Lent, Brown, & Hackett, 1994, 2000) recognize that early learning experiences are critical in the development of career interest, motivation, and choices. Learning experiences shape self-efficacy beliefs and outcome expectations, which in turn, affect the formation of vocational interests, which subsequently influence occupational goals, choice actions, and performance attainments. Thus, based on social cognitive career theories, we would expect that positive educational and learning experiences would shape self-confidence and career aspirations among students, including those on learning related IEPs.

According to a comprehensive literature review of 66 reports involving science education for students with disabilities (Mastropieri & Scruggs, 1992), knowledge and learning are facilitated through providing activities-oriented science curricula. Thus, we propose that the use of hands-on educational activities would lead to better learning for students on IEPs. In addition to better learning, such activities should lead to increased self-confidence and career motivation.

Pedagogical Approaches to STEM for This Population

Federal legislation mandates that academic accommodations be made to ensure that qualified postsecondary students with disabilities have educational opportunities that are equivalent to others. Faculty and staff members who are familiar with disabilities, accommodations strategies, and resources are better prepared to make arrangements that will ensure that students with disabilities have equal opportunities to participate in their programs. Educators who are familiar with universal design principles and accommodation strategies are better prepared to make arrangements that will ensure that students with CDs have equal opportunities to participate in their programs. We must continue to make faculty and staff members aware of: (a) the rights, potential contributions, and needs of students with disabilities; (b) educators' responsibilities for ensuring equal educational opportunities for all students in their programs; and (c) strategies for accommodating students who have disabilities in science and mathematics classes. Moreover, providing intensive training on weekends and during the summer involving hand-on exercises will lead to increased self-confidence and career interest in the areas of STEM among students with CDs.

Based on the review of the literature (Access STEM, 2007; Gosselin & Macklem-Hurst, 2002, Mastropieri & Scruggs, 1992; Norman, 1997), it appeared that the best approach to building interest and self-confidence was one that relied upon inclusive, inquiry-based science, emphasized problem-based learning, and incorporated visual demonstration. Group work and active learning based teaching have been proposed as effective practices for use with students and teachers in general, as well as students with disabilities (Access STEM 2007; Gosselin & Macklem-Hurst, 2002; Norman 1997).

Because of the diverse needs of students with CDs, we must always remember to incorporate principles of universal design in education (Access STEM, 2007; Dolan & Hall, 2001; Grumbine & Alden, 2006). The universal design of training or educational programs involves developing programs that are usable by all people, to the greatest extent possible, without the need for adaptation or specialized design (Burgstahler, 2006a, 2006b; Dolan & Hall, 2001). In creating course content, this involves paying attention to principles such as: (1) provide multiple media for the presentation of materials and deliver materials clearly and in multiple ways; (2) motivate all students; (3) design training to accommodate diverse learning styles; (4) use large, tactile aids; (5) provide cognitive and memory support including emphasis of major points and outlines; (6) make training practical, relevant, and hands-on; (7) facilitate interaction through group work; and (8) allow for peer interaction and feedback (Burgstahler, 2006a, 2006b).

In order for students with disabilities to pursue postsecondary academic and career options, they must have access to the high tech tools available to their nondisabled peers. These include computers, Web sites, Internet-based distance learning courses, instructional software, and scientific equipment. Achieving this goal requires that appropriate assistive technology be readily available and barriers to electronic tools and resources be eliminated. For example, it is important that students who are blind have access to speech and/or Braille output devices. But access to this assistive technology is not enough. In order for them to benefit fully from this technology, the educational software, applications software, Web pages, and other electronic resources they use must be designed in such a way that their full functionality can be accessed by using their keyboard and speech or Braille output system.

Legal mandates (e.g., Individuals with Disabilities Education Act, Section 504 of the Rehabilitation Act, and Americans with Disabilities Act) that apply to computer access for students and employees with disabilities are not always reflected in practice. Consumers and service providers identify the two biggest barriers to assistive technology access to be the lack of knowledge about appropriate assistive technology and lack of funding. Stakeholders are not fully aware of technology options, legal issues, and advocacy strategies. These stakeholders include people with disabilities, parents and mentors, government entities, paraprofessionals, policymakers and administrators, precollege and postsecondary educators, librarians, technical support staff, and employers. Studies have found that other access challenges include (a) lack of trained professionals to evaluate assistive technology, and (b) the bureaucracy of public programs and insurance companies (National Council on Disability, 2000; National Center for Educational Statistics, 2000). Educational systems need to overcome these challenges in order to ensure: (1) that people with disabilities gain access to the technology that has the potential to promote positive postsecondary and career outcomes, (2) that people with disabilities learn to use technology in ways that contribute to positive postsecondary academic and career outcomes, (3) a seamless transition of availability of technology as students move from K-12 to postsecondary and career environments, (4) the right balance between universal design and the provision of assistive technology in academic and employment computing environments, (5) policies and procedures should be established at all academic levels to ensure that universal accessibility is considered when electronic and information technology is procured, (6) policies, procedures, training, and support should be established at all educational levels to ensure that Web page, library resource, and distance learning program developers make their electronic resources accessible to everyone, (7) interagency collaboration on planning, funding, selecting, and supporting assistive technology should be fostered to ensure continuous technology access and support as students with disabilities transition through academic levels and to employment, (8) students with disabilities should be included at all stages of technology selection, support, and use, so that they learn to self-advocate regarding their needs for accessible technology in the classroom and workplace, (9) students with disabilities at high school and college levels should participate in internships and other work-based learning experiences where they can practice using technology in work settings, and (10) Legislators and policy makers should disseminate information about current laws, policies, and resources that are universally designed to meet the needs of various stakeholders. They should also identify and correct inconsistencies and gaps in legislation and policies regarding the selection, funding, and support of technology for people with disabilities.

More specifically, access to electronic and information technology has the potential to promote positive postsecondary academic and career outcomes for students with disabilities. This potential will not be realized, however, unless stakeholders (a) become more knowledgeable about appropriate uses of technology, (b) secure funding, and (c) work together to maximize the independence participation and productivity of students with disabilities as they transition to college, careers, and self-determined lives. Ultimately, ensuring that all of the educational and employment opportunities that technology provides are accessible to everyone will strengthen our economy and contribute to the creation of a level playing field.

K-12 Teachers' Strategies and Resources for Working with This Population in STEM

During the 2008-09 school year, approximately 87% of children with speech or language impairments were served in general education classrooms, 6% in resource rooms, and 4.6% in separate classes (U.S. Department of Education, 2010b). A wide variety of service delivery models for students with CDs are used within and across these three educational placement options. ASHA recognizes the following seven service delivery models (ASHA, 2000).

Monitoring: The SLP monitors or checks on the student's speech and language performance in the general education classroom. This option is often used just before a student is dismissed from therapy.

Pull-Out: The traditional and still most prevalent model of service delivery is the pull-out approach, sometimes called *intermittent direct service*. SLPs spend two-thirds of their time working with a child individually or with small groups of up to three children (ASHA, 2010). Depending on the needs of the individual child, pull-out may involve sessions of up to 1 hour 5 days per week.

Collaborative Consultation: Increasingly, communication disorders specialists serve as consultants for general and special education teachers (and parents) rather than spending most of their time providing direct services to individual children (Dohan & Schulz, 1998). SLPs who work in school settings more often as team members concerned with children's overall education and development. The SLP often provides training and consultation for the general education classroom teacher, who may do much of the direct work with a child with communication disorders. The specialist concentrates on assessing CDs, evaluating progress, and providing materials and techniques. Teachers and parents are encouraged to follow the specialist's guidelines.

Separate Classroom: Students with the most severe CDs are served in special classrooms for children with speech or language impairments. During the 2009-10 school year, approximately 1 to 20 children with speech or language impairments were served in separate classes (U.S. Department of Education, 2011).

Community Based: In community-based models, speech and language therapy is provided outside the school, usually in the home. This model is most often used with preschoolers and sometimes for students with severe disabilities, with an emphasis on teaching functional communication skills in the community.

Combination: Variations of all these models exist, and many schools and SLPs serve children using combinations of two or more models.

Classroom or Curriculum Based

Increasingly, SLPs are working as educational partners in the classroom, mediating between students' communication needs and the communication demands of the academic curriculum. SLPs report devoting about one-fourth of time in helping teachers integrate language and speech goals into daily curriculum activities. The advantage is that services are brought to the child and the teacher, and communication connections with the curriculum are made more directly.

In fact, educational planning for children with speech and language impairments involves many factors. Planning concepts that need to be considered in the classroom setting include seating arrangements, reducing distractions in the physical environment, and interactive techniques to enhance the teaching learning process. For children with central auditory processing disorders, parents, general and special education teachers, and therapists should focus on training certain auditory and listening skills, such as auditory discrimination (for example, telling the difference between peas and bees), localization of sound, sequencing sounds, or identifying a target sound in a noisy background. Training these skills in isolation, however, may not help a child understand complex language, such as a teacher's instructions. Therefore, another approach concentrates on teaching more functional language skills (vocabulary, grammar, conversational skills) and uses strategies (visual aids, repeating directions) to facilitate the processing of language.

Changes at home and in the classroom can also help a child with central auditory processing problems.

Seating: To help the child focus and maintain attention, select seating that is away from auditory and visual distractions. A seat close to the teacher and the blackboard and away from the window and the door may be helpful.

Setting: Reduce external visual and auditory distractions. A large display of posters or cluttered bulletin boards can be distracting. A study carrel in the room may help. Earplugs may be useful to block distracting noise from a heater or air conditioner, the pencil sharpener, or talking the hallway. Check with an audiologist to find out if earplugs are appropriate and which kind to use. Placing mats and cloth poster boards on classroom walls has been shown to decrease the reverberation of noise. A structured classroom setting may be more beneficial than an open classroom situation.

To improve the listening environment, an audiologist may recommend the use of a device that transmits the teacher's voice directly to the student's ear while blocking out background noise. The audiologist can provide recommendations on the potential benefit of available options based on the child's individual needs. As a key member of a child's IEP team, the SLPs goal is to correct the child's speech and/or language problems or to help the child achieve the maximum communicative potential, which may involve compensatory techniques and/or augmentative and alternative means of communication. Speech-language pathology addresses both organic and functional causes and encompasses practitioners with numerous points of view who use a wide range of accepted intervention techniques.

Some SLPs employ structured exercises and drills to correct speech sounds; others emphasize speech production in natural language contexts. Some prefer to work with children in individual therapy sessions; others believe that group sessions are advantageous for language modeling and peer support. Some encourage children to imitate the therapist's speech; others have the child listen to recordings of his own speech. Some specialists follow a structured, teacher-directed approach in which targeted speech and language behaviors are precisely prompted, reinforced, and recorded; others use less structured methods. Some SLPs focus on a child's expressive and receptive communication; others devote attention to other aspects of the child's behavior and environment, such as developing self-confidence and improving interactions with parents and classmates. A general goal of specialists in CDs is to help the child speak as clearly and pleasantly as possible so that a listener's attention will focus on the child's message rather than how he says it.

Differentiating STEM Instruction for This Population of Learners

Although students with CDs are often characterized as having mild to severe difficulties, including them in general education classrooms can become very complex and demanding, even for skilled teachers. Moreover, at a time when many school systems around the world have embraced the inclusionary model to serve a diverse student body, it is vital that instructors have the necessary tools to teach STEM in the inclusionary classroom. Little attention to diverse learners is given in content area courses because STEM instructors may not have the tools that allow them to meet the needs of the academically diverse students in their classroom.

Nevertheless, when congress reauthorized the Individuals with Disabilities Education Improvement Act (IDEIA), one of the most significant changes was the stipulation that the special education and related services prescribed in a child's IEP be supported by empirical research to the extent practical."

Thus, Congress made a legal requirement of something many special educators had always strived to do: use the results of scientific research to ensure their students receive the highest-quality instruction. It is unfortunate that a federal law is required to motivate educators to use scientifically sound teaching practices with children whose learning is most dependent upon effective instruction.

The reality, however, is that far too few teachers use evidence-based practices in their classrooms (Burns & Ysseldyke, 2009), many students with disabilities have been the recipients of teaching methods that are misguided at best, and some students have been subjected to practices that research has shown repeatedly to be not only benignly ineffective but also harmful (Heward, 2003; Jacobson, Foxx, & Mulick, 2005).

In reality, students want to learn and their educators, share this goal. The question remains, how can STEM educators design their instruction to maximize the learning of all students? The field of universal design for learning (UDL) can provide a starting point. UDL is an approach to designing the environment and products that takes into consideration the wide variety of characteristics of individuals and the changes experienced by people during their lifetime.

Rather than focusing on adapting things for an individual at a later time, an accessible product, activity or environment is created from the beginning. Disability is just one of many characteristics considered. For example, one person could be five feet four inches tall, female, forty years old, a poor reader, and deaf. All of these characteristics, including her deafness, should be considered along with those of other people when developing environments, products, or services.

Some of these suggestions (e.g., incorporating visual, aural, and tactile demonstrations in instruction) benefit all students, not just those with specific disabilities. UDL is an approach to teaching that involves consideration of students with a wide range of characteristics, including disabilities.

First applied in the field of architecture, UDL is the creation of products and environments for the use of all individuals devoid of adaption or specialized design (The Center for Universal Design, 2014).

UDL can be classified at three levels: (1) research based practice, (2) promising practices, and (3) emerging practice. Many students with disabilities face challenges to gaining knowledge. Examples of specific challenges and accommodations follow:

- The student who has difficulty: hearing presentations and instructions due to hearing impairments can be accommodated with: (a) FM system; (b) interpreter; (c) printed materials; (d) facing student for lip reading; and (e) overhead projector or blackboard.
- The student who has difficulty: hearing multimedia presentations due to hearing impairments can be accommodated with: (a) captioned presentations; and (b) sign language interpreter.
- The student who has difficulty: participating in class discussions due to hearing or speech impairment can be accommodated with: (a) electronic communications (e.g., email) where the ability to hear or speak is not required; and (b) portable computer with speech output.
- The student who has difficulty: doing research can be accommodated with: information accessible on computer (e.g., disk, Internet) with assistive technology.

Demonstrating Knowledge

Some students with disabilities cannot demonstrate mastery of a subject by writing, speaking, or by working through a problem in a lab. Many of the accommodations for gaining knowledge can also help the student demonstrate mastery of a subject. Examples of other accommodations follow.

- The student who has difficulty: completing a test or assignment because of a disability that affects the speed at which it can be completed can be accommodated with: (a) extra time or (b) alternative testing arrangements.
- The student who has difficulty: completing a test or assignment because of the inability to write can be accommodated with: in-class access to a computer with alternative input devices (e.g., Morse code, speech, alternative keyboard).

Although differentiated instruction is not a new idea, the differentiation movement has taken center stage as a means of meeting the needs of all students in the classroom. Differentiation involves finding multiple ways to structure a lesson so that each student has an opportunity to work at a moderately challenging level. It is an organized, yet flexible way of proactively adjusting teaching and learning to meet students where they are, while helping all students achieve maximum growth as learners (Tomlinson, 1999). Instruction may be differentiated in content, process, or product according to the students' readiness, interest, or learning profile. For example, all of the students may be studying force and motion (content), but the laboratory experiments in which they participate may be at varying levels of complexity to accommodate their academic readiness for a particular task (process).

Successful differentiation will occur in the classroom when a number of essential elements are also addressed. These essential elements include specific classroom management techniques that address the special needs of a differentiated STEM classroom, planned use of anchoring activities, a variety of differentiated instructional strategies, and differentiated assessment. More specifically, differentiated instruction refers to the use of slightly different lesson plans and approaches for different students. It may involve the use of extra accommodations: (1) additional resources, (2) extra help after class, and (3) different tests or extra time on tests.

CONCLUSION

There are many important reasons to support the need for STEM instruction for learners with communication disorders. K-12 teachers require pedagogical approaches, strategies and resources for working with this population in STEM. Differentiating STEM instruction for learners with communication disorders can be beneficial in supporting students in STEM education and the pursuit of STEM careers. The speech and language characteristics of students with communication disorders are significant factors that contribute to the academic and adjustment issues of many of these students. Therefore, general and special educators must design and implement instructional strategies that highlight strengths and areas of improvement for these students. For inclusive practices to be effective, in STEM, for students with communication disorders, each and every professional must have the best interest of the child in mind.

The active engagement of students with communication disorders in the learning process is critical. Thus, there are prescribed instructional models that aid in the planning, implementation, and evaluation of instructional STEM programs. General and special educators are urged to use differentiation, modification, group learning, hands-on learning, chaining, and natural selection or strategies and methods that fit each student that have been well-researched and documented as effective instructional strategies for improving the academic outcome of students with communication disorders in STEM.

REFERENCES

Access STEM. (2007). *Building capacity to include students with disabilities in science, technology, engineering, and mathematics fields*. Seattle, WA: University of Washington.

American Speech-Language-Hearing Association. (2010a). *Roles and responsibilities of speech-language pathologists in schools (Professional issues statement).* Rockville, MD: Author.

American Speech-Language-Hearing Association. (2000). *Guidelines for the roles and responsibilities of speech-language pathologists (Guidelines).* Rockville, MD: Author. Available from www.asha.org/policy.

American Speech-Language-Hearing Association. (1993). Definitions of communication disorders and variations. *ASHA 35* (Suppl. 10), 40-41.

Burgstahler, S. (2006a). *Equal access: Universal design of instruction.* Seattle: DO-IT, University of Washington.

Burgstahler, S. (2006b). *Universal design in education: Principles and applications.* Seattle: DO-IT, University of Washington.

Burns, M., & Ysseldyke, J. (2009). Reported prevalence of evidence-based instructional practices in special education. *Journal of Special Education, 43,* 3-11.

Dolan, R. P., & Hall, T. E. (2001). Universal design for learning: Implications for large-scale assessment. *IDA Perspectives,* 27, 22-25.

Gosselin, D. C., & Macklem-Hurst, J. L. (2002). Pre-/post knowledge assessment of an earth science course for elementary/middle school education majors. *Journal of Geoscience education*, 50, 169-175.

Grumbine, R., & Alden, P. B. (2006). Teaching science to students with learning disabilities. *Science Teacher*, *73*(3), 26-31.

Hale, J. E. (2001). *Learning while black: Creating educational excellence for African American children.* Baltimore: Johns Hopkins University Press.

Heward, W. L. (2003). Ten faculty notions about teaching and learning that hinder the effectiveness of special education. *Journal of Special Education, 36(4)*, 186-205.

Individuals with Disabilities Education Act. (IDEA, 1975). Formerly called P. L. 94-142 or the Education for all Handicapped Children Act of 1975.

Jacobson, J. W., Foxx, R. M., & Mulick, J. A. (Eds), (2005a). *Controversial therapies for Developmental disabilities: Fads, fashion, and science in professional practice.*Mahwah, NJ: Eribaum.

Lent, R. W., Brown, S. D., & Hackett, G. (1994). Toward a unifying social cognitive theory of career and academic interest, choice, and performance. *Journal of Vocational Behavior, 45*, 79-122, 1994.

Lent, R. W., Brown, S. D., & Hackett, G. (2000). Contextual supports and barriers to career choice: A social cognitive analysis. *Journal of Counseling Psychology,* 47, 36-49.

Mastropieri, M. A., & Scruggs, T. E. (1992). Science for students with disabilities, *Review of Educational Research,* 62, 377-411.

National Center for Educational Statistics. (2000). *What are the barriers to the use of Advanced telecommunications for students with disabilities in public schools?* (Rep. No. 2000-042). Washington, DC: U.S. Department of Education, Office of Educational Research and Improvement.

National Council on Disability and Social Security Administration. (2000). *Transition and post-school outcomes for youth with disabilities: Closing the gaps to postsecondary education and employment.* Washington, DC: Author.

National Council on Disability. (2000). *Federal policy barriers to assistive technology.* Washington, DC: Author.

National Science Foundation. (2000). *Women, minorities, and persons with disabilities in science and engineering.* Washington, DC: Author.

National Joint Committee on Learning Disabilities. (NJCLD, 1998). *Operationalizing the NJCLD definition of learning disabilities for ongoing assessment in schools.* Retrieved from http://www.Idonline.org/about/partners/njcld

Norman, K. I. (1997). A university grant project on science education for students with disabilities: Teaching elementary science inclusive classrooms. *1997 AETS Conference Proceedings.* Retrieved from www.ed.psu.edu/ci/Journal/97pap34.htm.

Tomlinson, C. A. (1999). *The differentiated classroom: Responding to the needs of all learners.* Alexandria, VA: Association for Supervision and Curriculum Development.

U.S. Department of Education. (2011). *Individuals with Disabilities Education Act (IDEA) data (Table 1-3).* Washington, DC: Author. Retrieved from https://www.ideadata.org/arc_toc11.asp#partbCC.

U.S. Department of Education. (2010b). *Individuals with Disabilities Education Act (IDEA) data (Table 2-2i).* Washington, DC: Author.

In: S.T.E.M. Education
Editor: Satasha L. Green
ISBN: 978-1-63117-806-1

Chapter 3

WORKING WITH LEARNERS WITH COGNITIVE DISABILITIES IN STEM

Sunday O. Obi*
Kentucky State University, US

ABSTRACT

Many students with cognitive disabilities face challenges to acquiring skills or gaining knowledge. Some students with disabilities have conditions that are invisible; some are visible. As scientific fields make increasing use of Science, Technology, Engineering, Mathematics (STEM) content, new opportunities emerge for students with a broad range of abilities.

Moreover, when students with disabilities and STEM teachers form learning partnerships, the possibilities for academic and career success multiply. More specifically, when the student and teacher work together to develop creative alternatives (accommodations) for challenges faced by students with cognitive disabilities, their outcomes improve.

The purpose of this chapter is to provide K-12 general and special education teachers, practices and strategies in STEM content to support the learning needs of students with cognitive disabilities. This chapter, will discuss (a) briefly cognitive disabilities and prevalence; (b) the importance of STEM instruction for learners with cognitive disabilities; (c) the pedagogical approaches to STEM for this population; (d) K-12 teachers' strategies and resources for working with this population in STEM; and (e) differentiating STEM instruction for this population of learners.

INTRODUCTION

The concept of cognitive disabilities is extremely, broad, and not always well-defined. The cognitive processes were believed to take place within the central nervous system (CNS) and include such things as language, memory, attention, and perception (Hammill, 1993;

* E-mail: Sunday.obi@kysu.edu.

Moats & Lyon, 1993). A person with a cognitive disability has greater difficulty with one or more types of mental tasks than the average person. There are many types of cognitive disabilities. These include Autism, Down Syndrome, Traumatic Brain Injury (TBI), Attention Deficit Disorders (ADD), Dyslexia, Dyscalculia, and Learning Disabilities in general. A person with profound cognitive disabilities will need assistance with nearly every aspect of daily living.

On the other hand, a person with a minor cognitive disability may be able to function adequately despite the disability, perhaps even to the extent that the disability is never discovered or diagnosed. Some of the main categories of cognitive disabilities include difficulties with: (a) memory, (b), attention, (c) metacognitive deficit, (d), generalization, (e) reading, linguistic, and verbal comprehension, (f) visual comprehension, (g) socio-emotional, (h) math comprehension, and (i) problem-solving.

Memory

Memory problems of students with cognitive disabilities may be manifested in learning simple material such as sight vocabulary or math facts, or may reflect a lack of effective strategies for learning more complex materials. Students with cognitive disabilities also frequently have difficulty with working memory, which is the memory that is needed to perform a particular task (Feifer, 2011).

Problems with working memory impact the ability to see something, think about it, and then act on this information (Siegel, 2003). The term "rehearsal" describes the unconscious cognitive strategies students use to remember. The strategy of rehearsal consists of using various verbal strategies, including repetition, rhyme, sub-vocal speech, and verbal self-instruction (Henley, Ramsey, & Algozzine, 1993).

Research has shown that students with mild cognitive disabilities are less likely to use spontaneous rehearsal procedures and are unable to benefit from incidental learning cues in their environment as compared to their non-disabled peers. Bray, Fletcher, and Turner (1997) presented evidence indicating that students with cognitive disabilities use rehearsal strategies if they comprehend the task demands. When these students are taught how to use rehearsal strategies, their short- and long-term memory skills improve (Borkowski & Day, 1987; Mercer & Snell, 1977).

Attention

Students with cognitive disabilities have difficulty distinguishing and paying attention to relevant cues in learning and social situations and attending to several different cues simultaneously (Zeaman & House, 1963, 1979). In addition, intelligence-related differences between students with cognitive disabilities and their non-disabled peers is the ability to perceive stimuli; this appears to be explained best in terms of physiological facts and differences in their abilities to orient toward, select, and group stimuli (Tomporowski & Tinsley, 1997). Finally, these students might have difficulty sitting in class and may be easily distracted.

METACOGNITIVE PROCESSES

Metacognitive processes, according to Sternberg and Spear (1985), are used to plan how to solve a problem, monitor one's solution strategy as it is being executed, and to evaluate the results of this strategy once it has been implemented (p. 303). Research has shown that students with cognitive disabilities do not develop efficient learning strategies or techniques that will facilitate the acquisition, manipulation, integration, storage, and retrieval of information across situations and settings.

Earlier research by Borkowski, Peck, and Damberg (1983) showed that students with cognitive disabilities are unable to find, monitor, or evaluate the best strategy to use when confronted with a new learning situation. In other words, these students do not approach learning strategically, depend upon others for learning, and do not generalize knowledge and skills learned in controlled settings to new situations and circumstances (Vaughn, Bos, & Lund, 1986). By contrast, their non-disabled peers use conspicuous learning strategies, actively respond to the learning demands of the classroom, and monitor task demands relative to other tasks (Palincsar, David, Winn, & Stevens, 1991).

GENERALIZATION

Generalization refers to the extent to which students with cognitive disabilities have extended what they have learned across settings and over time (Stokes & Baer, 1977; Stoks & Osnes, 1989). Research has shown that these students had impairments in their ability to transfer or generalize information and skills learned in one setting or one way to new situations which involves different expectations and skills (Langone, Clees, Oxford, Malone, & Ross, 1995).

READING, LINGUISTIC, AND VERBAL COMPREHENSION

Students with cognitive disabilities have difficulties understanding text. These difficulties may be mild or severe, ranging from minor challenges to a complete inability to read any text. Vocabulary knowledge is critical to academic success in the classroom. Individual differences in the language experiences and opportunities to learn new words vary immensely for students with cognitive disabilities. Research on the relevant learning characteristics of these students, as compared to their normal-achieving peers, has shown that these students differ in the association, comprehension, and generation of vocabulary words (Baumann & Kameenui, 1991; Westling, 1986). In addition, research has shown that for these students vocabulary problems are evident in terms of the number of words known and depth or knowledge, word learning is related to experience and exposure of words early in life, and the acquisition of vocabulary knowledge is very frustrating and intimidating for poor readers (Anderson & Nagy, 1991; Baumann & Kameenui, 1991). For students with cognitive disabilities, reading achievement levels are based upon their mental age. Although these students have impairments in their reading skills, they are typically better at decoding skills than reading comprehension.

Visual Comprehension

Some students with cognitive disabilities have difficulties processing visual information. In many ways, this is the opposite of the problem experienced by students with and verbal processing difficulties. Students with visual comprehension difficulties may not recognize objects for what they are. They may recognize the fact there are objects on a video, but may not be able to identify the objects. For example, they may not realize that a photograph of a person is a representation of a person, though they can plainly see the photograph itself (as an object) on the video. For these students, a moving, talking person in a video may be easier to identify and mentally processed than a static image of a person in a photograph. Video and multimedia, accompanied with narration, work effectively with these students.

Language

Language consists of a series of domains such as syntax, morphology, lexicon, phonology, speech acts, conversation, and discourse (Snow & Pan, 1993). Research on individual differences in language development in students with cognitive disabilities has shown that these students are relatively good at lexical, speech act and conversational skills but considerably impaired in morphology, syntax, and discourse skills (Snow & Pan, 1993). These researchers also suggest that some domains of language (e.g., speech acts, conversation, concept formation, and discourse) may be more sensitive to cultural and other environmental influences.

For students with cognitive disabilities, the way in which language is coded may impair how information is retrieved and stored in both long-term memory and working memory. There is clear evidence that languages input and affects the rate and qualify of language development for students with and without cognitive disabilities (Warren & Yoder, 1997). Kameenui and Carnine (1998) explained that language-based problems may be attributable to, in part, to poorly establish phonological features which aid in efficiently storing verbal information and the reliance on the semantic features or meaning of words rather than on the phonological features of words.

Social and Motivational Characteristics

Social skills are strength for many students with cognitive disabilities, but researchers estimate that approximately one third of these students have social-skills deficits (Lerner & Johns, 2009). These difficulties relate to how well students get along with teachers and peers. And these social-skills difficulties may lead to social isolation, become the target of bullies (Weiner, 2004) and motivational problems. It is important that teachers recognize and address these social issues, to ensure that they do not increase the students' adjustment problems and contribute to underachievement. Also, teachers must provide effective instruction that is appropriate to the students' needs, as well as teaching students learning strategies that they can use to actively engage in learning and by directly teaching social skills. Identifying the main categories of cognitive disabilities is more useful when considering STEM accessibility.

For example, telling a STEM instructor that some students have autism is not very meaningful unless the STEM instructor knows what kinds of barriers a student with autism might face.

However, telling an instructor that some students have difficulties comprehending math provides the teacher with a framework for intervening or teaching those students with comprehension problems. More specifically, STEM teachers must know what the student can or cannot do in order to properly work with the student.

IMPORTANCE OF STEM INSTRUCTION FOR LEARNERS WITH COGNITIVE DISABILITIES

There are many career opportunities that were once considered unachievable for individuals who have CD. However, with improvements in technology and more than before job specialization this population can work in STEM careers. Various STEM careers require attending postsecondary education to obtain the knowledge and skills to be successful. Even though individuals with disabilities in postsecondary education are increasing, the success rates of these students still lag far behind their non-disabled peers. In many demanding academic and STEM career fields persons with disabilities continue to be underrepresented. Therefore it becomes a challenge for Colleges and Universities to train an adequate number of qualified graduates in STEM fields. The supply and demand for individuals going into STEM areas is likely to continue to decrease, because of the reduction of the number of students pursuing degrees in STEM areas. However, by tapping into students from underrepresented groups in particularly students with CD, this may help to alleviate the potential shortages of individuals pursuing STEM careers (National Science Foundation, 2000).

One traditionally underrepresented group that has increased in size is students with cognitive disabilities (CD), including those students with learning disabilities [LD] (AccessSTEM, 2007; Grumbine & Alden, 2006; National Council on Disability and Social Security Administration, 2000; National Science Foundation, 2000). Students with LD have average to above average intelligence but have difficulties acquiring information and expressing their knowledge (IDEA, 1975; NJCLD, 1998). This may be reflected in speech, language, listening, speaking, writing, reasoning, or performing mathematical operations.

Schools are mandated by federal and state legislation to make programs accessible to individuals with disabilities e.g. Americans with Disabilities Act of 1990. Individuals with disabilities are disproportionately represented in many STEM careers. There are many factors that prevent individuals with disabilities from pursuing STEM education and careers: (1) negative stereotyping and (2) negative attitudes. Students with disabilities are often discouraged from taking math and science courses in middle and high school. When student with disabilities are enrolled in these classes they are often not fully included in the rigorous work required to be successful in these classes and beyond. Inadequate high school preparation restricts their options in college and careers.

Many individuals with disabilities are evaluated and defined by others solely based on their limitations and disabilities, rather than their capabilities, strengths, their broad range of interests and abilities. Many people assume that a disability limits an individual's ability to be educated. Therefore, educators must develop an attitude that is disability-positive and have high expectation for their students with disabilities. These students have the ability to be productive in the workforce. By targeting students in middle school and developing their interest, knowledge and confidence in STEM throughout high school may help to reduce the number of individuals going into STEM education and careers which can positively impact the labor shortages in these areas.

Social cognitive theories (Lent, Brown, & Hackett, 1994, 2000) acknowledge that early learning experiences are essential in the enhancement of career interests and motivation. Learning experiences form self-efficacy beliefs and outcome expectations, which in turn, affect the development of vocational interests, which influence occupational goals, choice actions, and performance attainments. Therefore, based on social cognitive career theories, positive educational and learning experiences can shape self confidence and career aspirations among all students. For example, science education for students with disabilities should be facilitated through providing activities-oriented science curricula (Mastropieri & Scruggs, 1992) and the use of hands-on educational activities. This should not only help to support increased learning, but such activities should lead to increased self confidence and career motivation.

Pedagogical Approaches to STEM for This Population

Federal legislation dictates that assurances are made to accommodate the needs of postsecondary students with disabilities so that they have equal educational opportunities as their non-disabled peers. Therefore it is imperative that educators are familiar with universal design for learning (UDL) principles and strategies to ensure that students with CD have opportunities to participate in STEM programs. Teachers and education professionals must understand: (a) the rights, potential contributions, and needs of students with disabilities; (b) educators' responsibilities for ensuring equal educational opportunities for all students in their programs; and (c) strategies for accommodating students who have disabilities in STEM classes.

Approaches that build interest and self-confidence rely upon inclusive, inquiry-based science, emphasized problem-based learning, and incorporated visual demonstration (AccessSTEM, 2007; Gosselin & Macklem-Hurst, 2002; Mastropieri & Scruggs, 1992; Norman, 1997). Other effective practices for students with disabilities are group work and active learning based teaching (Access STEM, 2007; Gosselin & Macklem-Hurst, 2002; Norman, 1997). UDL in education helps to address the diverse learning needs of students with cognitive disabilities (Access STEM, 2007; Dolan & Hall, 2001; Grumbine & Alden, 2006). UDL requires the development of programs that can be utilized by all people, to the greatest extent possible, without the need for adaptation or specialized design (Burgstahler, 2006a, 2006b; Dolan & Hall, 2001). This entails giving attention to principles such as : (1) providing multiple media for the presentation of material and delivering material clearly and in multiple ways; (2) motivating all students; (3) designing training to accommodate diverse learning styles; (4) using large, tactile aids; (5) providing cognitive and memory supports including emphasis of major points and outlines; (6) making training practical, relevant, and hands-on; (7) facilitating interaction through group work; and (8) allowing for peer interaction and feedback (Burgstahler, 2006a, 2006b).

K-12 Teachers' STEM Strategies and Resources

Pulling students aside for special instruction in the general classroom can be more stigmatizing than teaching them in a separate classroom. Resources of time, personnel, training, and materials are often inadequate, and in the absence of the needed supports, teachers often are ineffective. In short, there are many potential problems in teaching students with CD in inclusive settings, and research strongly suggests that poorly implemented or inappropriate inclusion frustrates teachers and short-changes students (Deno, Foegan, Robinson, & Espin, 1996; Weiss & Lloyd, 2003; Zimond & Baker, 1996).

Nevertheless, when inclusion is properly implemented for students with CD in STEM, teaching in inclusive settings can be highly rewarding for teachers and beneficial to students. The key is planning carefully, allocating adequate resources to support students with CD in STEM in an inclusive setting, and understanding that inclusion of these students in STEM programming is neither best for all students nor necessarily best done for the entire school day. Working out the specifics of co-teaching and other inclusionary models requires careful attention to the details of what is going to be taught, what specific roles each teacher will play, what materials will be used, and how teaching and learning will be evaluated. Students with CD respond positively to effective instructional principles. Teaching these students in inclusive settings can be successful and highly rewarding. However, success will not occur automatically for these students if the general education teacher is not culturally responsive.

Culturally responsive instruction is often synonymous with effective instruction for students with cognitive disabilities. Instruction should be the most important variable in determining where students with CD receive their education. The environment that provides the most appropriate instruction, or can be modified to provide that instruction, for each STEM student with CD is, in fact, the most appropriate placement, or least restrictive environment (LRE) – legal mandate for placement. Moreover, the No Child Left Behind Act (NCLB) requires students with disabilities to participate in the state accountability systems for math and reading in grades 3 through 8 and in science. The act does allow for accommodations to the standardized tests as long as the tests remain reliable and valid so that students with disabilities can be included in the testing data reported. Finally, NCLB requires that teachers use "scientifically-based instruction" in their classrooms (U.S. Department of Education, 2002a).

With the advent of ideas such as full inclusion, schools have restructured their service delivery models so that special educators move with their students into the general education environment, redefine professional roles, drop professional labels, and establish collaborative partnerships with other teachers (Thousand & Villa, 1989). These instructional partnerships between special and general education exist under the umbrella term *collaboration*. We can define collaboration as "the interaction that occurs between two professionals, often between the special education teacher and the general education teacher, and the roles that they play as equal partners in a problem solving endeavor" for students with disabilities (Box & Vaughn, 2002, p. 481).

Generally, two forms of instructional collaboration exist in the schools; (1) collaborative consultation and (2) co-teaching. Collaborative consultation involves the special educator acting as an expert consultant to the general educator, who is primarily responsible for the instruction of students with disabilities. The two share responsibility for the student and work together to meet the child's special education needs. The special educator, however, does not necessarily provide direct services to that child in the general education classroom. Co-teaching, on the other hand, puts the special educator into the general education classroom, working side by side with the general educator to provide instruction. According to the National Center on Educational Restructuring and Inclusion (1995), co-teaching is the collaborative model used most often by schools.

There are five basic models of co-teaching in the classroom (Vaughn, Schumm, & Arguelles, 1997). The models are flexible in that they are chosen to meet the needs of the students and the instructional task. The models are:

- *One teach-one assist* – Both teachers are present but one teacher takes the lead in delivering instruction. The other teacher then observes or moves around the room to monitor or assist students individually.
- *Station teaching* – Teachers divide the content to be delivered, and each takes responsibility for teaching their part to smaller groups of students who move between stations. Teachers divide students into three groups; two groups work with teachers, and one group works independently at all times.
- *Parallel teaching* – Teachers plan instruction together but split the class up and deliver the same instruction to smaller groups within the same classroom.
- *Alternative teaching* – One teacher works with a smaller group of students to re-teach, pre-teach, or supplement the instruction received by the larger group.
- *Team teaching* – Both teachers share the instruction of all students at the same time.

Additionally, team teaching can be broken into four parts:

1) *Tag team* – One teaches a part, and the other follows.
2) *Speak and add* – One teaches, one adds information.
3) *Speak and chart* – One teaches, one records on overhead, easel, etc.
4) *Duet* – Teachers work in unison, finishing each other's sentences and ideas. (Sands, Kozleski, and French, 2000).

To support the successful inclusion of students with CD in STEM programs, STEM instructors must adopt these strategies in their classrooms. Teachers must differentiate instruction by using technology to ensure that all students' needs are met. Students with disabilities need models and multisensory approaches tend to work best for students with CD.

Differentiating STEM Instruction for This Population of Learners

It is vital that instructors have the necessary tools to teach STEM in inclusionary classrooms. Students with CD are often characterized as having mild or moderate difficulties and including them in general education classroom can become complex and demanding. Many school systems around the world have embraced the inclusionary model to serve a diverse student body however little attention to diverse learners is given in STEM content area courses. Clearly, there is a need to provide teachers with tools that allow them to meet the needs of the academically diverse students in their classroom.

Differentiated instruction is one such philosophy that allows STEM teachers to respond to the needs of all learners. It is based on the idea that teachers need to adapt instruction to respond to student differences. STEM teachers may respond to differences in readiness, interest, and/or learning profile within the content, the process, or the product. In other words, differentiated instruction for students with CD in STEM may require teachers to use slightly different lesson plans and approaches for this population of students. Also, instructors might give students with CD in STEM extra accommodations: (a) additional resources, (b) extra help after class, and (c) different tests or extra time on tests. More specifically, differentiation involves finding multiple ways to structure a lesson so that each student has an opportunity to work at a moderately challenging level. It is an organized, yet flexible way of proactively adjusting teaching and learning to meet students where they are, while helping all students attain their maximum potential (Tomlinson, 1999). Students with CD respond positively when instruction is differentiated in content, process, or product according to the students' readiness, interest, or learning profile. For differentiation of instruction in STEM subjects to occur successfully, the teacher moves from being the one to impart knowledge to a facilitator of knowledge, and the student moves from being the consumer of knowledge to the producer of knowledge. Above all, flexible grouping arrangements create opportunities to meet individual needs, which is the basis for differentiated instruction. Students are rearranged for each lesson based on the lesson design and their individual needs. Groups can be formed in a variety of ways, including pairs, triads, or quads, as well as whole groups and small groups for instruction. Groups should be formed by both teachers and students, depending on the situation.

Effective instruction for students with CD in STEM allows a flexible use of time – lessons proceed to their natural conclusion rather than be carried out in set blocks of time. That is a lesson is not just something that is 30 minutes in duration; it may last for several days, where the richness of the activities dictates the amount of time spent on any individual day. To increase academic engagement, academic instruction must be adapted to meet the unique needs of individual students. STEM teachers of students with CD must also arrange the desks or tables so that various configurations can be made to facilitate group work, as well as whole class groupings that encourage sharing of ideas.

In order to effectively serve students with CD in STEM programs, instructors must devote the time, energy, and patience required to learn to effectively differentiate instruction for this population of students. In addition, teachers need support from administrators and their peers, as well as professional development over extended periods of time. In inclusive classroom settings students with and without CD represent a heterogeneous group of students. As a result, students vary in their instructional levels and teachers define goals and objectives and tailor instruction according to the learning and social needs of students. Also, when teachers decide how to teach, they must identify grouping arrangements, instructional approaches, and practices that move students toward their instructional goals and objectives. Lastly, effective instructional planning involves setting goals and expectations to the students so that they can understand what they are expected to do.

Another component of effective STEM instruction is managing instruction. It involves establishing classroom rules, teaching students to comply with rules, establishing and enforcing consequences, and handling disruptions, using time productively and establishing a positive classroom environment. That is, fostering acceptance of individual differences among peers and facilitating positive interactions among students. In fact, for teachers to be effective, they must continually make decisions concerning the: (a) presentation of information, (b) monitoring of information, and (c) use of adaptations and accommodations. Many strategies are focused on motivating students, providing opportunities to practice, giving immediate and explicit feedback, keeping students actively involved in the learning process, and adjusting lessons, the pace of instruction, and examining alternative instructional practices. The last component, evaluating instruction, is critical to knowing whether the instructional practices implemented in the classroom are effective with individual students. Evaluation procedures involve: (a) monitoring student performance, (b) charting academic progress of students, (c) making decisions about their student performance, and (d) deciding the instructional procedures that are effective with students (Utley & Obiakor, 2000).

Cognitive-behavioral intervention (i.e., self-recording and self-instruction) are effective ways to improve the classroom performance of students with cognitive disabilities. Components of this procedure, as described by Meichenbaum (1977), consist of: (a) an adult performing a task while describing aloud different steps he or she is taking; (b) a child performing the same task under the direction of an adult; (c) a child performing the same task alone while describing aloud the steps he or she is taking; (d) a child performing the task while whispering the steps to himself or herself; and (e) a child performing the task without verbalizing the steps aloud. Self-recording instructional procedures are very similar to self-instruction procedures in that an adult must model the technique. However, the two instructional procedures differ in that students, along with their teachers, address specific problem behaviors such as answering problems accurately. The student has the responsibility of deciding if the behavior has occurred and records the results. Cognitive-behavioral instructional procedures have been used to teach mathematics (Albion & Saltzberg, 1982; Leon & Pepe, 1983) to students with cognitive disabilities.

Peer tutoring programs, as reviewed in the literature, have been successfully implemented across a variety of student populations (e.g., at-risk; economically disadvantaged; and students with cognitive disabilities), students in the same grade (same-age), and older students with younger students (cross-age) (Osguthorpe & Scruggs, 1986). Although several peer tutoring programs have been developed and implemented in general and special education classrooms, one particular system, Class Wide Peer Tutoring (CWPT), as defined by Greenwood, Delquadri, and Carta (1997) and Utley, Mortweet, and Greenwood (1997) have emerged as a systematic instructional procedure using competing teams and game format. Earlier, Maheady, Saca, and Harper (1988) described other components of CWPT, as follows: (a) an explicit presentation format, (b) contingent point earning, (c) systematic error correction strategies, and (d) public posting of student performance. The benefits for students with CD are numerous: (a) tutoring allows teachers to supervise students in a time-efficient manner during instruction, (b) mastery learning occurs because practice and correction procedures are built within the system, (c) subject matter may be taught in primary/secondary languages, (d) second language acquisition is fostered through the context of talking to peers, (e) positive self-esteem and interpersonal attitudes are fostered, and (f) tutoring is a supplemental instructional activity to class instruction (Greenwood, Carta, Walker, Arreaga-Mayer, & Dinwiddie, 1988).

One of the most promising ways of encouraging appropriate interactions among students with cognitive delays and their peers is to provide tasks that require cooperative effort

(Hudson, 1989; Johnson & Johnson, 1986; Kagan, 1992) which is often utilized in STEM education.

The basic principles fundamental to cooperative learning include simultaneous interaction, positive interdependence, and individual accountability (Kagan, 1992). The most effective interactions within a cooperative context are designed to promote both personal and task-related contact, equal status among participants, opportunities for students to demonstrate skills that are contradictory to stereotypical beliefs and participate in activities that enhance peer relations among students (Johnson, Johnson, & Maruyama, 1983; Kagan, 1992). Teachers must structure classroom learning experiences to capitalize on goal interdependence between their heterogeneous groups of children. For student with cognitive delays, the use of cooperative structures motivates individuals to work harder because these students are working for the group and not working individually in a competitive situation (Kagan, 1992).

As indicated earlier, there is growing concern that the United States is not preparing a sufficient number of students, teachers, and professionals in the areas of STEM. The question remains, how can we create a safe and equitable learning environment for students with CD in STEM programs? As noted, Section 504 of the Rehabilitation Act requires that colleges make reasonable adjustments for students with disabilities in order to protect them against discrimination on the basis of their disabilities. These adjustments can take three general forms: "adaptations in the manner in which specific courses are conducted, the use of auxiliary equipment, and modifications in academic requirements" (Brinckerhoff, et.al, 1992, p. 421). Table 1 provides some examples of relatively common accommodations for students with CD.

In order to increase the number of students graduating in STEM programs in the United States, active engagement of students with CD in the learning process is critical. Therefore it is imperative for STEM teachers to utilize effective instructional principles for this population of students.

Table 1. Common Accommodations for Students with Cognitive Disabilities

Adjustments in course requirements and evaluation	• giving extra time on tests • allowing students to take exams in a distraction-free room • allowing students to take exams in a different format (e.g., substituting an oral exam for a written one)
Modifications in program requirements	• waiving or substituting certain requirements (e.g., a foreign language) • allowing students to take a lighter academic load
Auxiliary aids	• providing tape recordings of textbooks • providing access to a Kurzweil Reading Machine (a computer that scans text and converts it into auditory output) • recruiting and assigning volunteer note-takers for lectures

Conclusion

In this chapter, educators were exposed to the diverse educational needs of students with CD in order to succeed in the areas of science, technology, engineering, and mathematics (STEM). The task of providing effective instruction to this population is challenging. Educators need to consider the impact of cognitive, emotional, and behavioral functioning as well as meta-cognitive processes on learning and instruction. As facilitators of learning seek to improve educational practices in the areas of STEM in schools, they must address how to maximize the fullest potential of learners with cognitive disabilities.

Instructional practices may need to be adapted to more effectively serve this population, of students especially in the inclusive general education STEM classroom. There is significant research on how students with CD learn best. Moreover, when the teacher and student work together to develop creative alternatives (accommodations) for the challenges faced by the students with CD in STEM, their outcomes improve. It is time that students with CD receive an appropriate education that utilizes their strengths and talents. This may help to diminish the concern that the United States is not preparing a sufficient number of students, teachers and professionals in STEM.

References

Access STEM. (2007). *Building capacity to include students with disabilities in science, technology, engineering, and mathematics fields. Seattle*, WA: University of Washington.

Albion, F. M., & Salzberg, C. L. (1982). The effect of self-instruction on the rate of correct addition problems with mentally retarded children. *Education and Treatment of Children,* 5, 121-131.

Anderson-Inman, L., Paine, S. C., & Deutchman, L. (1984). Neatness counts: Effects on direct instruction and self-monitoring on the transfer of neat-paper skills to non-training settings. *Analysis and Intervention in Developmental Disabilities,* 4, 137-155.

Borkowski, J. G., & Day, J. (1987). *Cognition in special children: Comparative approaches to retardation, learning disabilities, and giftedness.* Norwood, NJ: Ablex.

Borkowski, J. G., & Varnhagan, C. W. (1984). Transfer of learning strategies: Contrast of self-instructional and traditional training formats with EMR children. *American Journal of Mental Deficiency,* 38, 369-379.

Box, C. S., & Vaughn, S. (2002). *Strategies for teaching students with learning and behavior* pro*blems* (5th ed.). Boston: Allyn & Bacon.

Bray, N. W., Fletcher, K. L., & Turner, L.A. (1997). Cognitive competencies and strategy use in individuals with mental retardation. In W. E. MacLean, Jr. (Ed.), *Ellis' handbook of mental deficiency, psychological theory, and research* (pp. 197-218). Mahwah, NJ: Erlbaum.

Brinckerhoff, L. C., Shaw, S. F., & McGuire, J. M. (1992). Promoting access, accommodations, and independence for college students with learning disabilities. *Journal of Learning Disabilities*, *25*(7), 417-429.

Bryan, T., Burstein, K., & Ergul, C. (2004). The social-emotional side of learning disabilities: A science-based presentation of the state of the art. *Learning Disability Quarterly, 27*(1), 45-51.

Burgstahler, S. (2006a). *Equal access: Universal design of instruction.* Seattle: DO-IT, University of Washington.

Burgstahler, S. (2006b). *Universal design in education: Principles and applications.* Seattle: DO-IT, University of Washington.

Deno, S. L., Foegen, A., Robinson, S., & Espin, C. (1996). Commentary: Facing the realities of inclusion for students with mild disabilities. *Journal of Special Education*, 30, 345-357.

Dolan, R. P., & Hallo, T. E. (2001). Universal design for learning: Implications for large-scale assessment. *IDA Perspectives*, 27, 22-25.

Feifer, S. (2011). How SLD manifests in reading. In D. Flanagan & V. Alfonso (Eds.), *Essentials of specific learning disabilities identification* (pp. 21-41). Hoboken, NJ: Wiley.

Gosselin, D. C., & Macklem-Hurst, J. L. (2002). Pre-/post knowledge assessment of an earth science course for elementary/middle school education majors. *Journal of Geoscience Education*, 50, 169-175.

Greenwood, C. R., Carta, J. J., Walker, Arreaga-Mayer, C. A., & Dinwiddie, G. (1988). Peer tutoring: Special education. In T. N. Postlethwaite (Eds.), *The international encyclopedia of education: Research and studies* (Vol. 1, pp. 574-577), Oxford: Pergamon.

Greenwood, C. R., Delquadri, J. C., & Carta, J. J. (1997). *Together we can: Classwide peer tutoring to improve basic academic skills.* Longmont, CO: Sopris West.

Grumbine, R., & Alden, P. B. (2006). Teaching science to students with learning disabilities. *Science Teacher, 73*(3), 26-31.

Hammill, D. D. (1993). A brief look at the learning disabilities movement in the United States. *Journal of Learning Disabilities,* 26, 295-310.

Harris, K. R., Graham, S., & Pressley, M. (1992). Cognitive-behavioral approaches in reading and written language: Developing self-regulated learners. In N. N. Sighn & I. L. Beale (Eds.), *Current perspectives in learning disabilities: Nature, theory, and treatment* pp. 415-451). New York, NY: Springer-Verlag.

Hudson, P. J. (1989). Instructional collaboration: Creating the learning environment. In S. H. Fradd & M. J. Weismantel (Eds.). *Meeting the needs of culturally and linguistically different students: A handbook for educators* (pp. 106-129). Boston, MA: College-Hill.

Individuals with Disabilities Education Act. (1975). Formerly called P. L. 94-142 or the Education for all Handicapped Children Act of 1975.

Johnson, D. W., & Johnson, R. (1986). Mainstreaming and cooperative learning strategies. *Exceptional Children,* 52, 553-561.

Johnson, D. W., Johnson, R., & Maruyama, G. (1983). Interdependence and interpersonal attraction among heterogeneous and homogeneous individuals: A theoretical formulation and a meta-analysis of the research. *Review of Educational Research,* 53, 5-54.

Kagan, S. (1992). *Cooperative learning.* San Juan Capistrano, CA: Kagan Cooperative Learning.

Knapcyzk, D. R., & Livingston, G. (1973). Self-recording and student teacher supervision: Variables within a token economy. *Journal of Applied Behavior Analysis,* 6, 481-486.

Lent, R. W., Brown, S. D., & Hackett, G. (1994). Toward a unifying social cognitive theory of career and academic interest, choice, and performance. *Journal of Vocational Behavior,* 45, 79-122, 1994.

Lent, R. W., Brown, S. D., & Hackett, G. (2000). Contextual supports and barriers to career choice: A social cognitive analysis. *Journal of Counseling Psychology,* 47, 36-49.

Leon, J. A., & Pepe, H. J. (1983). Self-instructional training: Cognitive behavior modification for remediating arithmetic deficits. *Exceptional Children,* 50, 54-60.

Lerner, J., & Johns, B. (2009). *Learning disabilities and related mild disabilities* (11th ed.). Belmont, CA: Wadsworth.

Mastropieri, M. A., & Scruggs, T. E. (1992). Science for students with disabilities, *Review of Educational Research,* 62, 377-411.

Meichenbaum, D. (1977). *Cognitive-behavior modification.* New York: Plenum.

Mercer, C. D., & Snell, M. E. (1977). *Learning theory research in mental retardation: Implications for teaching.* New York, NY: Merrill/Macmillan.

Mercer, C., & Pullen, P. (2009). *Students with learning disabilities* (7th ed.). Upper Saddle River, NJ: Merrill/Pearson Education.

Moats, L. C., & Lyon, G. R. (1993). Learning disabilities in the United States: Advocacy, science, and the future of the field. *Journal of Learning Disabilities, 26, 282-294.*

National Center on Educational Restructuring and Inclusion. (1995). National study on inclusion: Overview and summary report. *National Center on Educational Restructuring and Inclusion Bulletin,* 2(2), 1-10.

National Science Foundation. (2000). *Women, minorities, and persons with disabilities in science and engineering.* Washington, DC: Author.

National Joint Committee on Learning Disabilities. (1998). Operationalizing the NJCLD definition of learning disabilities for ongoing assessment in schools. Retrieved from: http://www.ldonline.org/about/partners/njcld.

Norman, K. I. (1997). A university grant project on science education for students with disabilities: Teaching elementary science inclusive classrooms. Retrieved from:www.ed.psu.edu/ci/Journal/97pap34.htm.

Osborne, S. S., Kosiewica, M. M., Crumley, E. B., & Lee, C. (1987). Distractible students use self-monitoring. *Teaching Exceptional Children,* 19, 66-69.

Osguthorpe, R. T., & Scruggs, T., E. (1986). Special education students as tutors: A review and analysis. *Remedial and Special Education, 7*(4), 15-26.

Siegel, L. S. (2003). Basic cognitive processes and reading disabilities. In H. L. Swanson, K. R. Harris, & S. Graham (Eds.), *Handbook of learning disabilities* pp. 158-181). New York: Guilford.

Speece, D. L., & Cooper, D. H. (2002). A longitudinal analysis of the connection between oral language and orally reading. *Journal of Education Research*, 95, 259-272.

Thousand, J. S., & Villa, R. A. (1989). Enhancing success in heterogeneous schools. In S. Stainback, W. Stainback, & M. Forest (Eds.), *Enhancing all students in the mainstream of regular education* (pp. 89-103). New York: Brookes.

Tomlinson, C. A. (1999). *The differentiated classroom: Responding to the needs of all learners.* Alexandria, VA: Association for Supervision and Curriculum Development.

U.S. Department of Education. (2002a). *The No Child Left Behind Act Title I: Improving the academic achievement of the disadvantaged – Summary of final regulations.* Retrieved from http://www.ed.gov/PressReleases/11-2002/regs_sum.html

Utley, C. A., Mortweet, S. L., & Greenwood, C. R. (1997). Peer-mediated instruction and interventions. *Focus on Exceptional Children, 29*(5), 1-24.

Weiss, M. P., & Lloyd, J. W. (2003). Conditions for co-teaching. Lessons from a case study. *Teacher Education and Special Education,* 26, 27-41.

Weiner, J. (2004). Do peer relationships foster behavioral adjustment in children with learning disabilities? *Earning Disability Quarterly*, *27*(1), 21-30.

Zagmond, N., & Baker, J. M. (1996). Full inclusion for students with learning disabilities: Too much of a good thing? *Theory into Practice*, *35*(1), 26-34.

Zeaman, D., & Houise, B. J. (1963). The role of attention in retardate discrimination learning. In N. R. Ellis (Ed.), *Handbook of mental deficiency* (pp. 159-223). New York: McGraw-Hill.

Zeaman, D., & House, B. J. (1979). A review of attention theory. In N. R. Ellis (Ed.). *Handbook of mental deficiency: Psychological theory and research* (2nd ed.) (pp. 31-51). Hillsdale, NJ: Erlbaum.

In: S.T.E.M. Education
Editor: Satasha L. Green
ISBN: 978-1-63117-806-1

Chapter 4

WORKING WITH LEARNERS WITH LEARNING DISABILITIES IN STEM: IMPLICATIONS FOR A SUCCESSFUL MODEL

Audrey M. Sorrells**, Heather A. Cole, Barbara Pazey and Jessica F. Carter
University of Texas at Austin, US

ABSTRACT

Improving student achievement in science, technology, engineering, and mathematics (STEM) subjects has become a national priority since the start of the 21st century.

As the global economy becomes more dependent on technological industries, the need for a workforce that is skilled in science and math related fields is rising. In order to meet the workforce needs, many countries, including the United States, have started to promote STEM education initiatives as a way to ensure future financial stability. Contemporary STEM initiatives and programs are aimed at increasing student achievement in math and science in order to prepare them for careers in technology based fields; however, not all students are showing gains in STEM subjects.

Students with learning disabilities have yet to show significant achievement gains in math and science subjects, which is problematic considering that this population of students is an untapped talent pool that can contribute to the STEM workforce. Students with learning disabilities can face significant barriers to learning STEM subjects, but when effective instructional strategies are employed by teachers, they can achieve at the same levels as students without disabilities.

This chapter provides research-based instructional frameworks and strategies that can be used in STEM classrooms and provides an example of a successful school model that is helping students with disabilities excel in STEM subjects.

* E-mail: audrey.sorrells@austin.utexas.edu.

INTRODUCTION

The subjects of science, technology, engineering and mathematics (STEM), which, in recent years, has evolved into a significant part of K-12 education. STEM education "represents a symbiotic relationship between four interwoven fields" (Basham & Marino, 2013, p. 9). Beyond the content of the individual subjects, STEM education emphasizes a hands-on and collaborative approach to learning that helps students develop analytic skills that will be transferable in the workforce. Currently only 16 percent of American high school seniors are proficient in mathematics and interested in a STEM career (U.S. Department of Education, 2013).

President Obama articulated a clear priority for STEM education with a declaration that within a decade, American students must improve in science and math and progress to the top in these areas. Specifically, he has called on the nation to increase the number of highly qualified STEM teachers and increase the number of college and university graduates with STEM majors, particularly Hispanics, African Americans, and other underrepresented groups in STEM fields-including women, first-generation Americans, and individuals with disabilities (U.S. Department of Education, 2013). Students with disabilities in STEM—and particularly students with learning disabilities, continue to experience disproportional underrepresentation in middle and high school STEM programs and careers. Although approximately 10% of the U.S. workforce is disabled, the National Science Foundation reports that people with disabilities hold only 5%-6% of the STEM field jobs (Burrelli, 2007).

Currently 2.4 million students are diagnosed with learning disabilities (LD) and receive special education services in our schools, representing 41% of all students receiving special education (IDEA, 2010).While students with LD continue to represent the largest group served by the Individuals with Disabilities Education Act (IDEA), these students are either underrepresented, or underperforming in middle and high school science and mathematics courses. As a result most students with LD are unprepared or underprepared to assume postsecondary coursework in these subjects or pursue STEM baccalaureate degrees or careers. Despite the challenges that students with LD face in STEM areas, they are capable of succeeding when their instructional needs are met. Schools across the country are implementing innovative programs and strategies to help students with disabilities excel in STEM fields. In this chapter, we discuss the need for STEM education for students with LD and describe strategies for overcoming barriers and providing accommodations for this population in STEM programs. We present an illustrative case of an innovative and inclusive STEM high school that utilizes project-based strategies to assist students with learning disabilities to access and succeed in STEM curricula and activities.

STEM EDUCATION IN U.S. SCHOOLS

Careers that require an applied understanding of STEM concepts are starting to replace traditional manufacturing jobs in the United States and many other countries (Basham & Marino, 2010). With manufacturing, and other production related industries, serving as many countries' driving economic forces, countries have started to invest heavily in STEM education. The newfound attention that STEM education has received in America since the start of the 21st century is a result of the country's position in the global economy being challenged by economically competitive countries that are producing workers who are competent in science and technology based fields. China, for example, is outpacing the United States in science, engineering, and technology when indicators such as research and development expenditures, size of the research workforce, and amount of technology exports are examined (National Science Board, 2010). Additionally, American students are being outperformed by students from other countries on international assessments of math and science competency. The Program for International Student Assessment (PISA) assesses students from the 34 countries that make up the Organization for Economic Cooperation and Development as well as partner countries and education systems that wish to participate in the triennial academic assessments.

In 2009, the United States ranked 17th in mathematics literacy and 25th in science literacy based on scores from tests that were administered to a representative sample of 15-year-old American students (Fleischman, Hopstock, Pelczar, & Shelley, 2010). Out-producing and outperforming other countries is not the only reason that STEM education has become so important. Science education is also a fundamental part of life, as 21st century citizens use science and technology in their day-to-day living now more than ever. The National Research Council (2006) asserted that most people in the United States lack the basic understanding of science that is needed to make informed decisions about the many scientific issues affecting their lives. America still has a long way to go in reclaiming its position as a world leader in science and technology advancement, and STEM education is how the country will get there.

The need to improve the 21st century skills of students has brought together schools, businesses, foundations, and government agencies to develop initiatives that support STEM learning for American students. In 2007, U.S. President George W. Bush signed the America Creating Opportunities to Meaningfully Promote Excellence in Technology, Education, and Science Act (COMPETES), which provided federal assistance to support advances in research and design programs, STEM education, and entrepreneurship. COMPETES was reauthorized in 2010 by President Obama and included provisions to increase student interest and participation in STEM, improve STEM literacy, employ proven strategies and methods for improving student learning and teaching in STEM, and provide curriculum support materials (COMPETES Act of 2010). The National Science Foundation and non-profits like Project Lead the Way have helped to advance the provisions of the COMPETES act by working with schools and communities to encourage interest in STEM education. Additionally, foundations and non-profit organizations have been instrumental in cultivating partnerships between schools and STEM related businesses to create pipelines to get students into science and technology related jobs.

STEM Education for Individuals with Learning Disabilities

The federal initiatives that have put billions of dollars into growing both K-12 and higher education STEM initiatives have been successful in adding more STEM teachers to the education workforce, funding the purchase of resources to aid in student learning, and establishing innovative science and technology based school and community programs;

however, even with these initiatives, not all students have been able to benefit from STEM education. Specifically, students with LD have continued to struggle to excel in STEM fields, largely due to the nature of STEM education which overlooks the needs of diverse types of learners in its complex design.

The federal definition of LD states, in part, that LD is "a disorder in one or more of the basic psychological processes involved in under-standing or in using language, spoken or written, that may manifest itself in an imperfect ability to listen, think, speak, read, write, spell, or do mathematical calculations" (34 Code of Federal Regulations §300.8 (c) (10)). Students with LD are likely to face challenges in STEM subjects that are either a direct result of the "root" problems that characterize LD or that are the result of maladaptive responses to the environmental demands in the face of those root problems (Brigham, Scruggs, & Mastropieri, 2011). Learning in STEM courses require a complex form of student engagement that is often based on inquiry as students pose questions, make predictions, conduct experiences, and apply what they learn across disciplines. Students with LD are often challenged by tasks that require highly developed literacy skills, information recall, prolonged attention, and motivation. For students with disabilities, the way in which STEM concepts are taught, not the concepts themselves, present several obstacles to learning that are addressed in this chapter.

STEM Access and Success for Students with Learning Disabilities

Barriers to learning do not have to keep students with LD from excelling in STEM fields. Individuals in STEM professions are often thought of as highly gifted individuals (and many of them are); however, some of the world's most notable scientists, mathematics, and engineers are also individuals with disabilities (Ludlow, 2013). Stephen Hawking, for example, has significant motor disabilities and is known around the world as a mathematician and astronomer. Carol Greider grew up living with the challenges of dyslexia, but discovered a key enzyme in cancer treatment for which she has a Nobel Prize in Medicine. These STEM pioneers, and numerous others, show that individuals with disabilities can, and do, succeed in science and math related fields; however, so few students with disabilities actually enter the STEM workforce. In 2010, the U.S. Census data indicates that more than 54 million Americans, nearly 19% of America's civilian non-institutionalized population, have some level of disability (U.S. Census Bureau, 2010). Numerous factors have contributed to the underemployment of individuals with disabilities in STEM fields, however, the success of individuals with disabilities in STEM fields depends heavily on the strength of the U.S. public education system to not only make STEM education accessible for all types of students, but to ensure that STEM instruction adequately prepares students to transition into degree programs at institutions of higher education (Leddy, 2010).

Providing individuals with disabilities access to STEM education and careers is necessary to ensure that they are as much a part of the workforce as individuals without disabilities. In recent years, countries around the world have moved towards academic models that not only emphasize STEM education, but that also educate all citizens in the same curriculum based on the idea that doing so will yield economic and social benefits for society as a whole

(Baker, 2009). In the United States, contemporary education policy has reflected this belief, as recent years have brought about legislation aimed at assisting individuals with disabilities gain access to STEM education and careers.

The alignment of No Child Left Behind Act (NCLB) with the Individuals with Disabilities Education Act (IDEA) raised academic standards for students with disabilities and promoted accountability for teachers and schools to ensure that students with disabilities attained levels of proficiency that was similar to their peers without disabilities (Brigham, Scruggs, & Mastropieri, 2011). One key provision of NCLB is that it requires all schools to test all students, including those with Individualized Education Programs (IEPs) and 504 plans. Schools, districts, and states are also required to disaggregate the test results for several groups, including students who are receiving services under IDEA, and report on the achievements of these students. The NCLB accountability provisions have brought attention to the significant gaps in achievement between students with disabilities and students without disabilities in all subjects. The achievement gaps in science and math related subjects are the widest, and show the importance of finding ways to enhance STEM education for students with disabilities.

Barriers to STEM Education for Students with Learning Disabilities

The economic growth and financial stability of the United States depends on the country's ability to produce the type of workforce that is capable of excelling in STEM fields and that will advance innovation and productivity on a global scale. Individuals with disabilities are an underutilized talent pool and should be a part of this workforce. They can, when the education that they receive enables them to do so. The delivery of STEM education to students with disabilities is paramount to getting them to become part of the STEM workforce, and although educating students with disabilities in STEM fields is still a relatively new endeavor, there is a growing body of research aimed at helping teachers to effectively deliver instruction to this population of students. Specifically, practices such as Universal Design for Learning (UDL) and project-based learning, that actively involve students in exploration of STEM concepts, have proven to be successful in raising achievement levels for students with LD.

Classrooms and Teachers. Students with disabilities are increasingly being mainstreamed into classrooms with their peers who are non-disabled. Although some general education classes have the support of special education teachers, most students with disabilities are taught by general education teachers. Teaching STEM content can be a highly involved process, and adapting instruction for students with disabilities can be challenging for general education teachers who do not possess sufficient knowledge or skills to make proper accommodations and modifications to instruction and assessments. Meeting the expectations of STEM education can be as challenging for teachers as it is for their students. Studies examining the self-efficacy of science teachers show that for both practicing and teachers-in-training, there is a belief that they are inadequately prepared to make appropriate instructional adjustments for students with disabilities (McGinnis & Stefanich, 2007).

General education teachers typically have little to no understanding of the nature of disabilities and how to tailor instruction to meet the needs of diverse types of learners; however, they can develop the skill set when given the necessary tools. A 2002 study examining the academic performance of 114 middle school students over the course of one school year found that when teachers received extensive training that focused on efficacious practices in inclusive science classrooms, students with disabilities passed the district science exam at the same rate as their general education peers (Cawley, Hayden, Cade, and Baker-Kroczynski, 2002). Collaborating and co-teaching with special education teachers can also help general education teachers in STEM classrooms. Collaboration leads to better generalization of instructional strategies, more communication about students' needs, and opportunities to develop and implement authentic, interdisciplinary STEM opportunities (Israel, Maynard, & Williamson, 2013).

Literacy. Students with learning disabilities often have difficulties with reading and vocabulary. Data from the National Assessment of Educational Progress [NAEP] (2009) showed that 62% of students with disabilities scored below the basic level in reading on an assessment that was given to a representative sample of 8th graders from across the country. The challenges that students face with literacy are intensified in other content courses, such as STEM subjects, where instruction is primarily based on complex expository textbooks and other print materials. Students with learning disabilities often lack the literacy skills necessary to learn effectively from text-based instruction, and commonly fail to master complex scientific vocabulary at high levels. Students with reading disabilities need additional supports if they are to engage in meaningful learning in STEM courses.

Teachers can help develop the reading comprehension skills of students with disabilities by applying the literacy framework of before, during, and after (BDF) for each component of literacy (Israel, Maynard, & Williamson, 2013). Before reading STEM texts, teachers can help their students to frontload important vocabulary and concepts vital to comprehension by pre-teaching essential new vocabulary, surveying the text, and engaging students in a quick write to share what they already know. During reading, students should be taught how to proceed through texts if they do not understand the vocabulary or concepts. This can be done by using varied text structures such as illustrations, graphic organizers, and embedded outlines to convey ideas to students or by encouraging students to paraphrase main ideas and important details from the text in their own words. After reading, students should integrate their newly acquired knowledge with their previous knowledge, which can be done by creating graphic organizers or recording notes in a learning log (Israel, Maynard, & Williamson, 2013).

Universal Design for Learning (UDL) is another framework that is gaining popularity in STEM instruction for its ability to help students with LD overcome literary barriers. UDL uses multiple means of representation, expression and action, and engagement to plan instruction for presumed and known levels of learner variability. UDL also specifies that curriculum, instruction, and related materials should provide multiple representations of key concepts, principles, and vocabulary in order to help meet the needs of all learners (CAST, 2011).

Relevance and Experience: Students with disabilities are prone to having negative attitudes towards STEM subjects. These attitudes develop as students progress in school and encounter increasingly complex instructional barriers that reduce their ability to access and comprehend scientific information (Lee & Erdogan, 2007). Making content relevant is one way to get students with disabilities interested in STEM subjects, and this can be done through the use of hands-on, project-based learning. Because students with learning disabilities are prone to having difficulties with textual and verbal methods of instruction, experience based methods of instruction can serve as a way to teach content, show relevance of STEM subjects, and make learning interesting for students. Multiple studies have shown that hands-on, activity-based instruction yields positive outcomes in general education settings that include students with LD as well as in special education settings that are dedicated to the instruction of students with disabilities (Brigham, Scruggs, & Mastropieri, 2011).

Use of Technology: Students with learning disabilities can become overwhelmed in STEM courses when tasked with complex cognitive tasks. In these instances, modern technology can be useful in adapting STEM curriculum and instruction for students with disabilities. Tools such as computers and tablets can help to make STEM content more cognitively accessible for students in a variety of ways. The use of technology to enhance instruction aligns with UDL principles and has been shown to act as a compensatory scaffold that supports students' STEM learning (Marino, Coyne, & Dunn, 2010). Technology can also play a significant role in project-based learning, as teachers are increasingly using technology to provide interactive experiences for students.

STEM instructional technologies can take on many forms such as simulations and mobile learning applications. Simulations allow learners to engage in STEM concepts virtually through simulated experiments where they can manipulate parameters, thus aiding in their understanding of abstract or complex concepts. Online simulation programs such as PhET and Gizmos are free and easily accessible for use by teachers and students. The rise in popularity of mobile technologies such as tables and smart phones has led to an increase in STEM related learning applications, or apps. Apps provide a wide range of learning experiences such as simulations, literacy supports, and games (Israel, Maynard, & Williamson, 2013).

Families: Parents are often left out of the recommendations for aiding students with disabilities achievement; however, collaboration with families and caregivers can be helpful in developing a shared understanding of STEM and its place in greater society. (Basham, Israel, & Maynard, 2010). By providing education on the benefits of STEM instruction and the resources to help their children learn, stakeholders can be highly effective in reinforcing what schools and community-based STEM programs teach their children. Caregivers can also help teachers by providing insight on child-specific topics such as interests and preferences that can be used to enhance instruction. In the remaining chapter, we present a local example of how one American high school has been redesigned to close the STEM achievement gap among underrepresented groups by improving STEM opportunities and success for all students, including those with learning disabilities.

Redesigning America's High Schools: A Local Example

President Obama's plans for educational reform have been extensive. His hope is that America will lead the world in college attainment by the end of the decade (U. S. Department of Education, 2013). With an increasingly competitive job market and an ever changing, technologically driven global economy, such a goal seems daunting. But, schools are changing and educators are beginning to rethink the way they provide instruction to their students. In 2013, President Obama launched the High School Redesign Initiative. The initiative supports competitive grants to local educational agencies (LEAs) that partner with higher education institutions and other entities, including businesses, so that schools have opportunities to provide academic concepts for their students to real world challenges (U. S. Department of Education, 2013). The initiative cites eight critical areas in which schools should rethink how they address their students' learning experiences. In brief, those eight areas include: (1) redesigning academic content and instructional practices to better align with postsecondary education and careers; (2) personalize learning opportunities to support the educational needs and interests of individual students; (3) provide academic and wrap-around support services for students who need them, including students with disabilities; (4) provide high-quality career and college exploration and counseling for students on postsecondary educational options; (5) offer opportunities to earn postsecondary credit while still in high school; (6) provide career-related experiences or competencies; (7) strategically use learning time in more meaningful ways and; (8) provide evidence-based professional development to deepen educators' skills. In the same government fact sheet that lays out President Obama's High School Redesign Initiative, five U.S. high schools are noted as transformative high schools. These schools are designed to incorporate innovative methods that lead to increased academic outcomes, student engagement, and future student academic success (U. S. Department of Education, 2013). The high school that tops the list is Manor New Tech High School located just outside Austin, Texas. Manor New Tech High School is part of the New Tech Network, a network of 100 schools across 18 states that rely on project-based, student-centered learning models.

Modeled after Sacramento New Tech High School, which was modeled from the first New Technology High School in Napa, California, Manor New Tech's goal is "to prepare students to excel in an information-based and technologically-advanced society" (Manor New Tech High School website, 2013). The school focuses on STEM instruction and incorporates project-based learning, infused with technology as a central element in teaching and learning. The school provides every student with a personal iPad as well as the use of a desktop computer within each classroom, and access to a network that not only allows for but, depends upon collaboration with peers, businesses and the community. Mastering academic content is achieved through hands-on, authentic learning. This in turn fosters and develops students' critical thinking, problem solving, communications and creativity. According to Manor New Tech's website (2013), the school's curriculum "brings together the strength of modern technology, community partnerships, problem solving, interdisciplinary instruction, and global perspectives in a student-centered, collaborative, project-based community".

Manor New Tech opened its doors in 2007. It was founded in collaboration with the Texas High School Project, the Texas Governor's Office, Samsung Semi-Conductor and a $4 million grant through the Texas STEM grant program. When the school began, only 160 students were in attendance. In 2012, the school enrolled 345 students. Sixty- five percent of the students were below the poverty level. Seventy-five percent represented racial minorities. Eleven percent of the student body was identified with disabilities. Statistically, these are the students with the worst outcomes in public education. However, Manor New Tech students are succeeding. Leveraging their supporters, Manor New Tech gives their students opportunities to earn college credit, visit local laboratories and workplaces and be mentored by business and community leaders. The school boasts a 97 percent attendance rate, 0 percent dropout rate and 100 percent completion rate. For any school, these are impressive numbers. In addition, in the most recent graduating class of 2012, 97 percent were college/university bound with 82 percent of the students headed to four-year higher education institutions. Sixty-five percent of the graduating class was first generation college/university students (personal communication, principal, May 17, 2012).

Implementing Project-Based Learning

Manor New Tech prides itself on being a 100 percent project-based learning school. Project-based learning is a "systematic teaching method that engages students in learning knowledge and skills through an extended inquiry process structured around complex, authentic questions, and carefully designed products and tasks" (Buck Institute for Education, 2003, p. 4).

Project-based learning allows flexibility for teachers to differentiate their instruction to meet a variety of student needs. Because focus is on an overarching theme or problem, learning can take place across disciplines. Students are able to make sense of content and see how things come together as a whole. Students are not just allowed but encouraged to work together and to build upon one another's strengths so that every student can participate.

At the launch of a new project, students are given a scenario for the project along with specific content standards that the project must address. Students are also provided with initial project materials, timelines and calendars, and assessment rubrics. Students are then asked to work either individually or within their groups to determine what they know and what they need to know. These know/need to know lists are then discussed in the classroom setting so questions can be asked, clarifications made and all students can feel comfortable about the scope and expectations of the project. Student groups then apply their content knowledge, acquired across disciplines, to carry out the projects. Every student group has a designated leader but all students in the group are expected to participate. Each student receives a work ethic grade designed to foster responsibility for learning in each individual student (Yanez, Schneider, & Leach, 2009).

Student Interaction: Embracing Inclusive Classrooms

Manor New Tech students are accepted through a blind lottery. As such, their student body is diverse and students with disabilities make up a representative portion of their enrollment. Manor New Tech's approach to special education is full inclusion. With the project-based learning model, special education students are seamlessly integrated into the general education population. Students still receive the necessary accommodations dictated by their IEPs but this is accomplished within the greater school community, as special education students are not set apart but integral to the diversity so integral to the New Tech model. In practice, the school provides accommodations; however, the curriculum content remains the same; thus, there are no modifications in the expectations or demands the students are expected to achieve. One-on-one and small group instruction is arranged as a tutoring session, before-after school or during lunch. The project-based curriculum gives teachers the ability to offer students multiple ways of learning new concepts. Because instruction is spread across a number of disciplines, students can apply learning to a real, tangible problem. As Manor New Tech's website (2013) explains, "the project-based learning environment sets up an atmosphere where learning is student-driven, engaging, and meets the needs of a wide variety of academic abilities".

The use of technology gives teachers a variety of means to transmit information. Direct instruction is only a small piece. Students also learn with visual, audio and practical experiences. Because of the smaller, group-oriented approaches, teachers have additional time to provide individual support to any student who may need additional assistance or clarification on a particular skill. Prior to taking this route, however, students must exhaust their own resources and efforts, as a group, before seeking help from the teacher. The group projects also create peer support opportunities. Part of the work ethic grade is dependent upon how students work together and address both the strengths and challenges of their colleagues (Yanez, Schneider, & Leach, 2009).

Manor New Tech's principal, Steve Zipkes, in an on-line blog, advocated for the student-centered approach that is so integral to project-based learning and a full inclusion model. Students at Manor New Tech are encouraged to move within the classroom, think independently and collaboratively as well as use technology daily with all teachers in every subject. Manor New Tech has an inclusive model of education. In a television interview with a student from Manor New Tech who was diagnosed with dyslexia, admitted that he was not sure whether he would ever graduate from high school. He has been accepted to attend college and credits Manor New Tech with changing his attitude and his opportunities. In his own words, he told reporters "after coming to Manor New Tech, I learned how to embrace my disability and face it head on instead of running away from it or using it as a crutch" (Wiggins, 2013). The project-based learning model gave him the opportunity to work with peers and to not just visualize, but actually understand what he could contribute and where he needed additional help.

Implications for Administrators, Teachers and Students

A STEM model like at Manor New Tech requires a new way of thinking about all of a school's key players: administrators, teachers and students. Administrators need to have the flexibility to think outside the box and fashion a curriculum that is not solely test-driven. Accountability standards can make things tough when more and more instructional days must be given over to test taking.

For principals like Steve Zipkes, the key to a successful STEM school is relationships. This is not something that is discussed often in education reform circles where measures of academic proficiency and achievement are predicated by a test score and a rigorous (yet inflexible) curriculum. Manor New Tech doesn't just depend on test scores to evaluate students. Student achievement is also evaluated through portfolios of work and a comprehensive understanding of each student's work so that parents and teachers can get a true picture of a student's ability and can tailor each student's academic program. This can hardly be achieved by the more traditional approaches of rote instruction and a standardized test score.

Integrated classrooms, shared curriculum and group work allow teachers to foster better relationships with students with LD. Students can see that teachers encourage their independence and creativity and want them to begin to function on their own. Students are given the chance to do their own work as teachers guide and facilitate their progress. For teachers, this means training to allow them to be comfortable in this role, time to meet, share and collaborate with their teacher colleagues, and ongoing professional development to give them the tools they need to stay up to date with STEM instructional strategies and new technology. Teachers also need opportunity to be creative with the design of their projects so they can keep in mind state standards but also give their students a chance to learn with projects that are meaningful, interesting and have real world applications.

For students with LD, STEM education that works does not isolate them but brings them together and teaches them both independence but also collaboration and teamwork. Students experience a school where relationships with administrators and teachers are built on trust and respect and where their relationships with each other mirror that. In schools with such positive cultures, all students are welcome and all students can benefit. Students with LD are no longer the outliers. They are part of the team and their chances of succeeding once they leave high school are greatly increased. Students with LD should have equal opportunity to avail themselves of the benefits of a good STEM education. To repeat the words of Manor New Tech's principal, "we need to provide this opportunity for all kids" (Statesman, 2012). Indeed, given the country's desire for global competitiveness, it is the least we can do. Every child should be able to contribute to society upon graduation. The more we can give them, the greater that contribution will be.

CONCLUSION

The future of the United States is heavily dependent on the country's ability to produce a workforce that is capable of competing globally in STEM fields. Providing American students with a high quality STEM education is the only way to ensure that the country remains a leading global competitor. Students with LD need a high quality STEM education, and with appropriate instruction, they can achieve at the same levels as their peers without disabilities. There is a need to implement strategies and models that can lead to improving inclusive STEM programs in classrooms and schools around the country.

REFERENCES

America Creating Opportunities to Meaningfully Promote Excellence in Technology, *Education, and Science Act of 2010*, USC §5116.

Baker, D. P. (2009). The invisible hand of world education culture: Thoughts for policy makers. In G. Sykes, B.L. Schneider, D.N. Plank, & T.G. Ford (Eds.), *Handbook of education policy research* (pp.958-968). New York; Washington, DC: Routledge; American Educational Research Association.

Basham, J. D., & Marino, M.T. (2013). Understanding STEM education and supporting students through universal design for learning. *Teaching Exceptional Children, 45*(4), 8-15.

Basham, J. D., Israel, M., & Maynard, K. (2010). An ecological model of STEM education: Operationalizing STEM for all. *Journal of Special Education, 25*(3), 9 – 19.

Brigham, F. J., Scruggs, T. E., & Mastropieri, M. A. (2011). Science education students with learning disabilities. *Learning Disabilities Research & Practice, 26*(4), 223-232.

Buck Institute for Education. (2003). *Project based learning handbook: A guide to standards-focused project based learning for middle and high school teachers*. Novato, CA: Author.

Burrelli, J. S. (2007). *What the data show about how students with disabilities in STEM*. National Science Foundation Committee on Equal Opportunities in Science and Engineering (CEOSE).

CAST. (2011). *Universal design for learning guidelines* (Version 2.0). Wakefield, MA: Author. Retrieved from http://www.udlcenter.org/aboutudl/udlguidelines.

Cawley, J., Hayden, S., Cade, E., & Baker-Kroczynski, S. (2002). Including students with disabilities into the general education science classroom. *Exceptional Children*, 68, 423-435.

Fleischman, H. L., Hopstock, P. J., Pelczar, M. P., & Shelley, B. E. (2010). Highlights from PISA 2009: *Performance of U.S. 15-Year-Old Students in Reading, Mathematics, and Science Literacy in an International Context* (NCES 2011-004). U.S. Department of Education, National Center for Education Statistics. Washington, DC: U.S. Government Printing Office.

IDEA Part B Child Count. (2010). *Students ages 6-21*. Retrieved from http://www.IDEAdata.org

Israel, M., Maynard, K., & Williamson, P. (2013). Promoting literacy-embedded, authentic STEM instruction for students with disabilities and other struggling learners. *Teaching Exceptional Children, 45*(4), 18-25.

Khadaroo, S. T. (2013). *In Texas, Obama lauds 'New Tech' high school: Model for the future?* Retrieved from http:// www.csmonitor.com.

Leddy, M. H. (2010). Technology to advance high school and undergraduate students with disabilities in science, technology, engineering, and mathematics. *Journal of Special Education Technology, 25*(3), 3-8.

Lee, M., & Erdogan, I. (2007). The effect of science-technology-society teaching on students' attitudes toward science and certain aspects of creativity. *International Journal of Science Education*, 11, 1315-1327.

Ludlow, B. (2013). Growing stem as a pathway to the future. *Teaching Exceptional Children, 45*(4), 4.

Manor New Tech High School website. (2013). Retrieved from http://mnths.manorisd.net/apps/pages.

Marino, M. T., Coyne, M. D., & Dunn, M. W. (2010). Technology-based curricula: How altered readability levels affect struggling readers' passage comprehension. *Journal of Computing in Mathematics and Science Teaching, 29*(1), 31-49.

McGinnis, J. R., & Stefanich, G. P. (2007). Special needs and talents in science learning. In S. K. Abell & N. G. Lederman (Eds.), *Handbook of research on science education* (pp 287-317). Mahwah, NJ: Erlbaum.

Mini-Symposium: Students with Disabilities in STEM. Arlington, VA: *National Science Foundation.* Retrieved from http://www.nsf.gov/od/iia/activities/ceose/mini-symppres/Students_with_disabilities_in_STEM_Joan_Burrelli.pdf.

National Assessment of Educational Progress. (2009). *The Nation's Report Card Science.* Retrieved from http://nces.ed.gov/nationsreportcard/naepdata/dataset.aspx.

National Research Council. (2006). *America's Lab Report: Investigations in high school science. committee on high school science laboratories: Role and vision*, In S. R. Singer, M. L. Hilton, & H. A. Schweingruber, (Eds). Board on Science Education, Center for Education. Division of Behavioral and Social Sciences and Education. Washington, DC: The National Academies Press.

National Science Board. (2010). *Globalization of science and engineering research: A companion to science and engineering indicators 2010 (NSB 10-3).* Retrieved from http://www.nsf.gov/statistics/nsb1003/.

U. S. Department of Education. (2013). *Fact sheet: Redesigning America's high schools.* Retrieved from http://www.ed.gov/news/press-releases/fact-sheet-redesigning-americas-high schools.

U.S. Census Bureau. (2010). 20th Anniversary of Americans with Disabilities Act: July 26 (CB10-FF.13). *U.S. Census Bureau News: Facts for Features.* Retrieved from http://www.census.gov/newsroom/releases/archives/facts_for_features_special_editions/cb10-ff13.html.

Wiggins, M. (2013). *Manor students give President Obama warm welcome.* Retrieved from http://www.kvue.com/news/Manor-tech-school-awaits-President-Obamas-arrival-206779851.html.

Yanez, D., Schneider, C. L., & Leach, L. F. (2009). *Summary of selected finding from a case study of Manor New Technology High School, 2008-2009.* Austin, TX: Charles A. Dana Center at The University of Texas at Austin.

Zipkes, S. (2012). *Driving question: New thinking for new schools for the 21st century, what will it take?* Retrieved from http://www.p21.org/tools-and-resources/p21blog/1053-zipkes-blog.

In: S.T.E.M. Education
Editor: Satasha L. Green

ISBN: 978-1-63117-806-1

Chapter 5

IMPROVING THE OUTCOMES FOR STUDENTS WITH EMOTIONAL/BEHAVIORAL DISORDERS IN STEM

***Lenwood Gibson*[*1] *and Festus E. Obiakor*[2]**
[1] Queens College of New York, CUNY, US
[2] Valdosta State University, US

ABSTRACT

Students with emotional and behavioral disorders (EBD) have many difficulties in and out of the classroom and some of these difficulties translate into being unsuccessful in STEM education. There are many benefits for students who are successful in STEM disciplines including better chances for employment and higher paying jobs. Unfortunately, many students with EBD do not exhibit the prerequisite behaviors or academic skills to perform well in the classroom. Fortunately, there are many evidence based strategies to help teach these behaviors and skills to students with EBD. This chapter will discuss the use of positive behavior interventions and supports that help to improve behaviors of EBD students. Self-monitoring and self-regulated strategies will be discussed. Teachers and service providers need to continue to develop and investigate strategies that allow students with EBD to access and be successful in STEM disciplines.

INTRODUCTION

As the nation moves towards the implementation of educational programs for the future, the focus on science, technology, engineering, and mathematics (STEM) education has become an integral part of most K-12 curriculum. STEM education is a vital way of providing grade school students the foundation needed to be successful in the areas of science, technology, engineering, and mathematics. The benefits of such educational programs are numerous but a few include being better prepared for 21st century jobs, post-secondary learning, and competing in the global job market. In fact, students graduating from strong

[*] Lenwood.gibson@qc.cuny.edu

undergraduate STEM programs are less likely to be unemployed and more likely to earn higher incomes than students who graduate from non-STEM programs.

Although there are very positive benefits of STEM education for successful students, far too many students do not perform well within the areas that comprise STEM education. The latest report from the National Assessment for Educational Progress (NAEP, 2011) indicated that many students scored well below proficient levels in the areas of math and science. In 2011, only 35% and 32% of eighth grade students scored in the proficient range for math and science, respectively. Similar outcomes were also noted for 4th graders in both math and science education. These two areas are essential to the development of both engineering and technology so students struggling with science and math are also likely to struggle with these areas. Additionally, while the data from the Nation's report card points to the need to focus on improving outcomes for all students in the key STEM areas, students with disabilities are at a greater risk for struggling in science and math education. According to NAEP (2011), only 17% and 9% respectively of fourth and eighth graders with disabilities scored at or above proficient levels in math. Even worse, only 11% of eighth grade students with disabilities scored at or above proficient levels in science.

For many students with emotional and behavioral disorders (EBD), there are significant struggles with their behaviors that affect their learning. According to the Individuals with Disabilities Education Improvement Act (IDEIA, 2004), emotional disturbance is defined as a long lasting condition in which students' emotions and/or behaviors adversely affect their educational performance. Approximately 8.1% of students who receive services under IDEIA are categorized as EBD, but there are likely more unidentified students who are affected by emotional and/or behavioral problems (Cartledge, Gardner, & Ford, 2009). Although most teachers tend to focus on the social deficits of students with EBD, the initial identifying criterion of these students is "an inability to learn that cannot be explained by intellectual, sensory, or health factors" (IDEIA, 2004). This criterion specifically focuses on the learning process and how students with EBD struggle with the academics of the classroom. These struggles are present without any evidence of intellectual deficits or other factors. Apparently, there is a clear link between students with behavior problems and academic failure in the classroom (Algozzine, Wang, & Violette, 2010; Anderson, Kutash, & Duchnowski, 2001; McLeod & Kaiser, 2004; Payne, Marks, & Bogan, 2007).

McLeod and Kaiser (2004) conducted a study that demonstrated that students identified with EBD in early school years were significantly less likely to receive their high school diploma and enroll in college. This study used data to track a cohort of students from different socioeconomic backgrounds from elementary through high school. Some conclusions from this study indicate that students with EBD experienced academic difficulties throughout their school years. These difficulties were compounded as students moved further along in school and some students with EBD were retained in middle or early high school. The results of this study support the need to provide strong academic support for students with EBD throughout their schooling years but in particularly in the early grades.

Intertwined with issues of STEM and EBD is the disproportional representation of culturally and linguistically diverse (CLD) learners. For instance, there is evidence to suggest that African American students are 1.5-5.5 times more likely to be categorized at having EBD than other racial groups (Coutinho, Oswald, & Forness, 2002; Lo & Cartledge, 2006). Although the reasons for this overrepresentation are debatable, there is no argument on the effect that it has on these students. As previously stated, students identified with EBD have a marked inability to learn and they often struggle in the classroom with academic subjects. The combined effects of disproportional representation and the academic struggles exhibited by students with EBD place CLD students at a much higher risk of failure within STEM subjects. For example, when data from the NAEP are examined for African American and students with disabilities, these risk factors are confirmed. In the 2011 report, African American students in eighth grade performed well below basic and proficient levels in math and science. In fact, only 13% of African American students performed at or above proficient levels in math versus 49% performing below basic levels. The results for African American eighth graders in science were worse with only 10% performing at or above proficient levels and 63% performing below basic levels. Even as grim as these results are, African American students with disabilities are likely to have been performing even worse; however the NAEP did not report data for students with disabilities along racial lines.

Considering that many school programs moving towards curriculum that emphasize STEM education, it is critical to discuss ways to support students with EBD, and in particular CLD students with EBD. These students are especially vulnerable to academic difficulties due to their behavioral and/or emotional problems (Anderson, Kutash, & Duchnowski, 2001; Lane, Barton-Arwood, Nelson, & Wehby, 2008). The research literature is filled with studies to support a strong link between students with behavior disorders and academic failure (Anderson, Kutash, & Duchnowski, 2001; Lane, Barton-Arwood, Nelson, & Wehby, 2008). The purpose of this chapter is to discuss the impact that the difficulties faced by students with EBD have on their ability to be successful in STEM education. Embedded in this discussion are some evidence based solutions to increase the likelihood that these students will be successful with STEM education.

STEM Education for Students with Emotional and Behavioral Disorders

When teaching students with emotional/behavioral disorders several important factors should be taken into consideration. These students often exhibit specific maladaptive and disruptive behaviors that interfere with their academic performance. Behaviors that impact student learning can range from relatively mild (e.g., off-task or inattentiveness) to severe (e.g., aggression towards others). Furthermore, the reaction of the teacher or service provider to problematic behaviors can predict if those behaviors become better or worse. In order to help students with EBD improve their academic skills in STEM disciplines, teachers and other school professionals should employ evidence based strategies rather than non-researched approaches. In fact, there is growing support for the use of evidence based strategies when working with students with EBD (Niesyn, 2009; Ryan, Pierce, & Mooney, 2008). Some of these strategies focus on improving student behavior and tolerance to classroom expectations, including the use of using positive behavior interventions and supports (PBIS) and self-management strategies. Other strategies that directly address the academic skills needed to be successful in school and include improving academic vocabulary and using technology to differentiate instruction. When used effectively, these strategies can improve the academic performance of students with EBD and increase their chances to be successful in STEM domains.

Using Positive Behavior Supports

The use of positive behavior supports for student with EBD is not new to special educators; however, these strategies are more recently being used in inclusive classrooms. The goal of PBIS is to establish constructive learning environments for all students by using behavioral principles of prediction and positive reinforcement. Strategies range from establishing routines and procedures to creating individualized behavior support plans. These strategies can be used with all students but are particularly effective for students with EBD. By effectively managing negative behaviors of students with EBD, they are more likely to successfully engage in academic tasks of the classroom.

Improving Self-Monitoring Skills

A key characteristic of students with EBD is their difficulty in monitor and managing their own behaviors. They are prone to being irritable and exhibiting outburst that can be disruptive to their learning and the learning of other students. By improving self-monitoring and management skills of students with EBD, they become more likely to engage in STEM academic activities and less likely to be disruptive. There are several different components that make up self-monitoring and these components can be explicitly taught to students (Menzies, Lane, & Lee, 2009). These components are examined in the following subsections.

Identify and Define the Target Behavior

The first step in using an effective self-monitoring program is to identify the behavior that needs to be changed. Typically, behaviors include skills needed to complete academic tasks such as staying on task or completing assignments. For students to be successful in the academic arena, they need to exhibit a number of prerequisite behaviors that support academics. Behaviors can be identified and defined before they can be corrected. Once a behavior is identified and defined, it must be determined if the deficit is due to an acquisition or performance issue.

Acquisition problems occur when a student has never learned the skill that is needed to be successful in the classroom. These can be particularly distressing to students with EBD because their behavioral outburst can be triggered by frustration. In order to prevent some of this frustration, acquisition problems need to be addressed by directly teaching the desired behavior to the student. This can be accomplished by modeling desired behaviors and having students practice them while giving feedback.

The other type of deficit is a performance problem in which the student knows how to perform the required skill but chooses not to. In these situations, students may likely lack the motivation to engage in the desired behaviors. To increase motivation, teachers must look for effective reinforcers that will support students engaging in desired behaviors. Whether it is an acquisition or performance deficit, once the behavior that needs to be changed is identified, it also needs to be accurately defined. By using an accurate definition, the behavior can be individualized for specific students. This individualization will help make the self-monitoring process more successful and more likely to have the desired impact on students.

Design Procedure and Self-Monitoring Form

The second step in the self-monitoring process is to create a simple but effective way for the student to track his/her performance during STEM subject periods. This can be accomplished by creating a self-monitoring form that the student can use throughout the day. This form needs to be easy for the student to understand and use. Some general guidelines include breaking the form into smaller sections, e.g. each STEM period during the school day. By doing so, the student will have multiple opportunities to monitor his/her behavior throughout STEM classes. Another guideline is to ensure that the monitoring form is age appropriate. This will help younger students to understand how to use it and prevent older students from feeling stigmatized.

Teach the Skills

One of the most important steps in the self-monitoring process is teaching students how to engage with the strategy. The best way to do this is to spend some time with the student to go over the behavior that is being worked on and how to use the actual self-monitoring form. The teacher can model the behavior to the student and show him/her how to correctly complete the form using a role-playing activity. After the teacher explains and demonstrates how the process works, time should be allotted for the student to practice. It is recommended that practice sessions are kept short and that direct feedback is given on the accuracy of determining if the behavior occurred and if it was correctly recorded. Once the student can accurately and appropriately engage in the self-monitoring process, spot checking from time to time is needed to ensure the student is still using the form as intended.

Monitor Progress

Once the self-monitoring process is in place and being used with fidelity, students' progress needs to be tracked and monitored. The whole purpose of the self-monitoring system is to improve student behavior clearly tracking progress is an integral part of the system. The best way to track progress is to collect data on occurrences of target behaviors. Using the monitoring form, data is converted to the total number of occurrences of the target behavior or to a percentage of intervals in which the behavior occurred. By making the monitoring form simple, the data conversion will also be straightforward and easy to interpret. These data are best interpreted by putting them on a graph to create a visual representation to the student's progress. Another feature of using visual representation of performance is the ability to show the graphs to the student. This allows students to actually see their performance for themselves which can further motivate them to reach their goals.

Maintenance and Follow-Up

The final step in the self-monitoring program is to use techniques that to help maintain students' behavior. By using a self-monitoring program, a student may become overly reliant on it to help maintain the desired behavior. Although this system may be needed initially, it is better for the student to internalize ways to maintain the desired behavior. Once the behavior has improved enough to no longer be a concern, the self-monitoring program can be faded out. The easiest way to accomplish this is to gradually increase the length of the intervals on the self-monitoring form. As noted earlier, students may need many opportunities throughout the day to assess their own behavior; however, when the behavior starts to improve, the number of opportunities can be decreased.

The above description of a self-monitoring procedure is a general overview of how specific steps can be implemented for students with EBD. Although this procedure is easy to implement and is based on evidence based strategies, it is also supported by research literature for students with EBD (Blood, Johnson, Ridenour, Simmons, & Crouch, 2011; Cancio, West, & Young, 2004; Gulchak, 2008; Lo & Cartledge, 2006). In the Cancio et al. (2004), a simple self-monitoring system was presented as a component of interventions to improve homework completion of students identified as having EBD. The results of this study demonstrated that following the implementation of the intervention, the completion and accuracy of math homework completed increased drastically. Additionally, all the students increased the overall math level when comparing pretest and posttest measures.

A study by Lo and Cartledge (2006) looked at off task behaviors of several African American elementary school students. As noted earlier, for students to be successful in any STEM academic task, they need to master certain prerequisite skills. One of these important skills is the ability to stay on-task for extended periods of time. Students who exhibit high rates of off task behaviors are more likely to fall further and further behind on academic work. Lo and Cartledge implemented an intervention package that included a self-monitoring procedure. Students were taught how to monitor their own task behavior throughout the course of selected academic activities by using a self-monitoring sheet and a prompting device. Results of this study indicated that the strategy was effective in reducing the off task behavior for all of the students. Furthermore, once the self-monitoring strategy was withdrawn in the maintenance phase of the study the students were still able to maintain low levels of off task behaviors.

Self-Regulated Strategies Development

Another approach for helping students increase academic performance while managing their own behaviors is called self-regulated strategy development (SRSD). This approach uses a set of specific steps that focus on teaching students to apply a task-specific strategy (e.g. a mnemonic) in conjunction with techniques to help manage behavior (Graham & Harris, 1999). For example, when trying to teach students about the order of operations in math, a common mnemonic is" Please Excuse My Dear Aunt Sally" (Parentheses, Exponents, Multiplication, Division, Addition, Subtraction). Although this task-specific strategy is fairly easy to teach and learn, some students may have difficulties due to their interfering behaviors. By teaching students to use the SRSD approach, they will be more likely to be successful in using the task specific strategy and in controlling any competing behaviors. There are six basic stages in using the SRSD approach, namely:

1) *Develop Background Knowledge.* Teachers can introduce the task specific strategy along with teaching the required skills to use it. For the math mnemonic, the teacher needs to be sure that students understand and can perform all operations. If students are missing the needed skills, they must be taught them prior to starting the SRSD approach.
2) *Discuss It.* Once teachers are confident their students have the appropriate background knowledge and prerequisite skills of the task-specific strategy, it should be introduced to students. During this stage, an explanation of how the task-specific

strategy works and when it can be applied is given. Teachers should be sure to get a commitment from students who are participating with this strategy. At this stage self-monitoring, instruction, and reinforcement components are also introduced to the students.

3) *Model It.* After students demonstrate an understanding of the specific task and self-monitoring strategies, the next stage is to actually model all of the components in action. For the use of the math mnemonic, the teacher can present several different math problems and demonstrate how to apply the steps in the mnemonic to solve the problem. The teacher also models the various self-monitoring strategies to ensure that students understand and use them adequately. The self-regulation strategies should include teaching students self-statements that will support their use of the task-specific strategy.
4) *Memorize It.* The next stage is to have students commit to the steps in the task-specific, self-monitoring strategies, and self-statements to memory. For the math mnemonic, students should be able to easily state the order of operations and use them without any prompting from the teacher. They should also state all of the steps in the self-monitoring technique and recite the supporting self-statements.
5) *Support It.* Teachers should provide the appropriate level of scaffolded supports to ensure that students are using all strategies correctly and effectively. Up front, the level of support should be high as students get adjusted to using the SRSD. This level of support can be gradually withdrawn as the student becomes more proficient in using the strategy and as his/her academic performance improves. Students should also be encouraged to convert self-statements when implementing strategies.
6) *Independent Performance.* On-going performance monitoring should continue once the teacher withdraws high levels of support. Teachers should consistently but periodically check students' performance to assess if academic goals are being met. Instructional decisions should be based on the outcomes of these assessment checks. The level of support should be adjusted accordingly.

The use of SRSD has proven effective in a number of research studies for students with academic and behavioral difficulties (Ennis, Jolivette, & Boden, 2013; Lane, Harris, Grahamm Driscoll, Sandmel, Murphy, & Schatschneider, 2011; Mason, Kubina, Valasa, & Cramer, 2010). Ennis et al. (2013) used SRSD to improve the writing skills of a small group of students in an alternative school setting for students with EBD. In this study, teachers taught students to use two task specific strategies (i.e., the STOP and DARE mnemonic) in tandem with the SRSD approach. They assessed participants on writing elements, quality of writing, and total written words.

The results of this study demonstrated that the SRSD procedure was more effective in increasing writing skills of treatment students compared to the students who did not receive the intervention. It should be noted that 40% of the students in this study were African American, demonstrating that the SRSD was effective for this population as well. Although the vast majority of the research on SRSD focuses on the writing process, this strategy can be applied to many other academic domains including, STEM content areas.

Developing Content Specific Academic Skills

After students gain the required skills to perform in the classroom, they may still face difficulties with the specific academic content to do well in STEM areas. Many students with EBD enter the classroom already below grade level in at least one academic area (McLeod & Kaiser, 2004; Trout, Nordness, Pierce, & Epstein, 2003). As they progress further in their school career the gap between these students and their peers continue to grow (Benner, Kutash, Nelson, & Fisher, 2013). By addressing specific academic areas it is possible to improve students' academic performance and close the achievement gap (Obiakor & Yawn, 2014). This is particularly important for students in STEM disciplines because both math and science have complicated material that needs to be mastered for success. Therefore, teachers should focus on improving content specific academic skills to increase chances that students with EBD will do well in STEM areas. There are several approaches to accomplish this task, but two of such approaches include improving STEM vocabulary and using technology to differentiate instruction.

Improving STEM Vocabulary

One reason why students may struggle in STEM content areas is because math and science use large amounts of content-specific vocabulary. This vocabulary can be extremely difficult to comprehend and make learning the content nearly impossible. Students need to develop strong content specific vocabulary in order to understand the complexities associated with math and science concepts. Researchers realized this need and have analyzed various strategies to determine the most effective techniques for students with disabilities (Beck, McKeown, & Kucan, 2000; Kim & Linan-Thompson, 2013; Pierce & Fountiane, 2009; Seifert & Espin, 2012).

Beck, McKeown and Kucan (2002) developed a three tier model to address the differences in types of vocabulary used by students. Vocabulary words on tier 1 are used in everyday interactions between students and adults. These are common words that generally do not need to be directly taught and are not needed to support specific academic content. According to this model, tier 2 words are vocabulary typically found across the academic curriculum. These words are needed to understand a wide range of academic text but are fairly common and usually do not need to be directly taught. The third tier contains vocabulary words that are highly content specific. These words are only used during specific academic instruction and need to be understood for students to comprehend the academic content being taught. In terms of STEM content areas, much of the vocabulary would be considered tier three. Considering that vocabulary in STEM content areas is highly content specific it needs to be directly taught, especially to struggling students. Sibold (2011) provided suggestions of evidence-based strategies to teach tier 3 vocabulary to students. One such strategy is to use a "talk-through" with read aloud activities in which students and teachers read a complex text assignment out loud. New content specific vocabulary words are emphasized during these readings. This strategy allows students to hear the new vocabulary, use it in context, and practice the words in a small group setting. The read alouds can be repeated several times to reinforce the meaning of selected vocabulary words. Another strategy that was presented by Sibold (2012) is the use of an academic vocabulary journal.

With this strategy, students are taught to keep a word bank that includes a list of new content specific vocabulary words. To improve their comprehension of these words, students write definitions and use the word in an original sentence. To further help students learn these new words, the journal can include pictures of the word and any ideas that are connected to the new vocabulary.

The third strategy that can be used to help students learn new vocabulary is called the PACE strategy. PACE stands for Prediction, Association, Verification, and Evaluation. Using this strategy, students are taught to read new content specific vocabulary words as it appears in the textbook. They then use context clues to predict the meaning of the word. Next, students make an association between the new word and their own mental images of the word. Third, they verify their prediction by looking up the definition of the word in the dictionary, and finally, they evaluate how close their prediction was to the actual definition of the word. Through this process, students learn how to become more independent learners and in addition this strategy can be used with self-regulation and self-monitoring strategies that were discussed earlier in the chapter.

The use of the three tier model can also be applied to content specific academics like math and science. For example, Pierce and Fontaine (2012) discussed ways to apply concepts introduced by Beck et al. (2002) to vocabulary instruction in mathematics. This instruction should focus on the ability of teachers to (a) provide explanations that are student-friendly as opposed to dictionary definitions and (b) have students engage in lively activities that allow them to process word meanings. Using such activities have proven to be effective in increasing the vocabulary of students significantly (Beck & McKeown, 2007).

When looking at math vocabulary that poses the greatest challenge to students on math tests, Beck and McKeown (2007) noted that these tests included unusual, specific math words and/or those with ambiguous meanings. Math specific vocabulary was placed into two categories: technical words and subtechnical words. Technical words have very specific meanings that are directly related to math. Subtechnical words have common meanings to many other academic subjects but when applied to math concepts these meanings become more ambiguous.

The math meanings of these words also need to be explicitly taught. Due to the fact that math vocabulary can pose a challenge to struggling students, these words need to be taught explicitly. Pierce and Fountaine (2012) provided several suggestions when teaching both technical and subtechnical math vocabulary. The first approach is to develop and use definitions that are student-friendly.

These definitions need to explain the meaning of vocabulary words in everyday language. It is also important to have students engage in activities that help develop deeper understanding of words as opposed to rote memory. The second approach to teaching math vocabulary is to engage students in lively activities that illustrate the meaning of words. For example, the teacher can use small group activities to have students practice solving word problems together. By having students work together and practice using the math vocabulary in context, they are more likely to gain a better understanding of words. By focusing on developing strong content specific vocabulary, teachers improve the chances for students to perform better in STEM disciplines.

Using Technology to Differentiate Instruction

A major issue when working with students that need additional instruction is the limited amount of resources that are available to provide supplemental instruction. This lack of resources can be insufficient funds to purchase highly effective curriculum; however, resources in terms of staffing can be harder to overcome. Considering that the use of small group or one-to-one instruction is a hallmark of the Response to Intervention (RTI) paradigm, the need for small student to teacher ratios is vital to the success of students in these programs. Breakdowns can occur when there is not enough staff to provide the high quality instruction needed for students to make progress in STEM academics. Part of the solution to this problem is to use technology to provide differentiated instruction in areas of math and science.

As the spread and quality of computer technology continues to advance, it penetrates further and further into the classroom. Over the past several years the use of computers to supplement classroom instruction is evident in the research literature (Aronin & Floyd, 2013; Billingsley, Scheuermann, & Webber, 2009; Haydon, Hawkins, Denune, Kimener, McCoy, & Basham, 2012). The success of computer technology is undeniable and the potential benefits can be enormous as in the use of smartphones, ipads, and tablets. Using such technology to assist and supplement high needs academic areas for students that are struggling seems logical but practical. Many programs are specifically designed to address academics in the STEM disciplines (Billingsley, Scheuermann, & Webber, 2009; Burns, Kanive, & DeGrande, 2012; Meyen & Greer, 2010; Nordness, Haverkost, & Volberding, 2011).

A recent study was conducted that used computer assisted instruction to improve the math skills for student with EBD (Billingsley et al., 2009). In this study the use of a math worksheet strategy was compared with to the use of an iPad math application. All of the participants either engaged with a math worksheet on one day and an iPad application on another. Participants were assessed on the number of basic math facts they got correct in one minute and on student engagement. Results indicated that when the iPad condition was implemented students were more engaged in their work and performed better on the math assessments. Although this is only one study, it lends support to the use of computer-assisted learning strategies to help improve math and science skills of students with EBD.

Conclusion

Students with EBD have many difficulties in and out of the classroom and some of these difficulties translate into academic failures. Some of these students might manage to keep up for a while, but eventually their deficits begin to show. This is even more evident when they are faced with difficult academic disciplines such as STEM education. There are many benefits for students who are successful in STEM disciplines including better chances for employment and higher paying jobs. Unfortunately, many students with EBD do not exhibit the prerequisite behaviors or academic skills to perform well in the classroom. Fortunately, there are many evidence based strategies to help teach these behaviors and skills to students with EBD.

Clearly, the use of positive behavior interventions and supports is well founded as a means to improve behaviors of troubled students. However, specific strategies such as self-monitoring and self-regulated strategy development get at the heart of issues that interfere with student progress. Both of these strategies focus on teaching students to change their own behaviors and therefore place an emphasis on self-control. Once students are able to monitor and control their own behaviors they are much more likely to engage in academic tasks needed to be successful in STEM disciplines. Even with the use of PBIS approaches, some students with EBD still struggle with the academic rigor of STEM subjects. Finally, these students need supplemental support that is directly linked to improving academic skills. Teachers and service providers need to continue to develop and investigate strategies that allow students with EBD to access and be successful in STEM disciplines. To a large degree, our schools communities, and nation will benefit from socially adapted, scientifically and technologically oriented productive youth.

REFERENCES

Algozzine, B., Wang, C., & Violette, A. S. (2011). Reexamining the relationship between academic achievement and social behavior. *Journal of Positive Behavior Interventions,* 13(1), 3-16.

Anderson, J. A., Kutash, K., & Duchnowski, A. J. (2001). A comparison of the academic progress of students with EBD and students with LD. *Journal of Emotional and Behavioral Disorders,* 9(2), 106-115

Aronin, S., & Floyd, K. K. (2013). Using an iPad in inclusive preschool classrooms to introduce STEM concepts. *Teaching Exceptional Children,* 45(4), 34-39.

Beck, I. L., McKeown, M. G., & Kucan, L. (2002). Bringing Words to Life: Robust Vocabulary Instruction. Solving Problems in the Teaching of Literacy. *Guilford Press:* New York.

Beck, I., & McKeown, M. G. (2007). Increasing young low-income children's oral Vocabulary repertoires through rich and focused instruction. *The Elementary School Journal,* 107(3), 251–271.

Benner, G. J., Kutash, K., Nelson, J. R., & Fisher, M. B. (2013). Closing the achievement gap of youth with emotional and behavioral disorders through multi-tiered systems of support. *Education & Treatment of Children,* 36(3), 15-29.

Billingsley, G., Scheuermann, B., & Webber, J. (2009). A comparison of three instructional methods for teaching math skills to secondary students with emotional/behavioral disorders. *Behavioral Disorders,* 35(1), 4-18.

Blood, E., Johnson, J. W., Ridenour, L., Simmons, K., & Crouch, S. (2011). Using an iPod touch to teach social and self-management skills to an elementary student with Emotional/Behavioral disorders. *Education & Treatment of Children,* 34(3), 299-321.

Burns, M. K., Kanive, R., & DeGrande, M. (2012). Effect of a computer-delivered math fact intervention as a supplemental intervention for math in third and fourth grades. *Remedial and Special Education,* 33(3), 184-191.

Cancio, E. J., West, R. P., & Young, K. R. (2004). Improving mathematics homework completion and accuracy of students with EBD through self-management and parent participation. *Journal of Emotional and Behavioral Disorders,* 12(1), 9-22.

Cartledge, G., Gardner, R., & Ford, D. Y, (2009). Diverse learners with exceptionalities: Culturally responsive teaching in the inclusive classroom. *Pearson: Upper Saddle River, NJ.*

Coutinho, M. J., Oswald, D. P., Best, A. M., & Forness, S. R. (2002). Gender and sociodemographic factors and the disproportionate identification of culturally and linguistically diverse students with emotional disturbance. *Behavioral Disorders,* 27(2), 109-125.

Ennis, R. P., Jolivette, K., & Boden, L. J. (2013). STOP and DARE: Self-regulated strategy development for persuasive writing with elementary students with E/BD in a residential facility. *Education & Treatment of Children,* 36(3), 81-99.

Graham, S., & Harris, K. R. (1999). Assessment and intervention in overcoming writing difficulties: an illustration from the self-regulated strategy development model. *Language, Speech & Hearing Services In Schools,* 30(3), 255-264.

Haydon, T., Hawkins, R., Denune, H., Kimener, L., McCoy, D., & Basham, J. (2012). A comparison of iPads and worksheets on math skills of high school students with emotional disturbance. *Behavioral Disorders,* 37(4), 232-243.

Individuals with Disabilities Education Improvement Act (IDEIA) of 2004, PL108-446, 20 USC §§ 1400 et seq.

Kim, W., & Linan-Thompson, S. (2013). The effects of self-regulation on science vocabulary acquisition of English language learners with learning difficulties. *Remedial & Special Education,* 34(4), 225-236.

Lane, K. L., Barton-Arwood, S., Nelson, J. R., & Wehby, J. (2008). Academic performance of students with emotional and behavioral disorders served in a self-contained setting. *Journal of Behavioral Education,* 17(1), 43-62.

Lane, K. L., Harris, K., Graham, S., Driscoll, S., Sandmel, K., Murphy, P., & Schatschneider, C. (2011). Self-regulated strategy development at tier 2 for second-grade students with writing and behavioral difficulties: A randomized controlled trial. *Journal of Research on Educational Effectiveness,* 4(4), 322-353.

Lo, Y., & Cartledge, G. (2006). FBA and BIP: Increasing the behavior adjustment of African American boys in schools. *Behavioral Disorders,* 31(2), 147-161.

Mason, L. H., Kubina, R., Valasa, L. L., & Cramer, A. (2010). Evaluating effective writing instruction for adolescent students in an emotional and behavior support setting. *Behavioral Disorders,* 35(2), 140-156.

McLeod, J. D., & Kaiser, K. (2004). Childhood emotional and behavioral problems and educational attainment. *American Sociological Review,* 69(5), 636-658.

Menzies, H. M., Lane, K. L., & Lee, J. M. (2009). Self-monitoring strategies for use in the classroom: A promising practice to support productive behavior for students with emotional or behavioral disorders. *Beyond Behavior,* 18(2), 27-35.

Meyen, E. L., & Greer, D. L. (2010). Applying technology to enhance STEM achievement for students with disabilities: The blending assessment with instruction program. *Journal of Special Education Technology,* 25(3), 49-63.

National Center for Education Statistics. (2011). The nation's report card: Science and math. National Assessment of Educational Progress at Grades 4 and 8. NCES 2010-458. National Center for Education Statistics, Retrieved from EBSCOhost.

Niesyn, M. E. (2009). Strategies for success: Evidence-based instructional practices for students with emotional and behavioral disorders. *Preventing School Failure,* 53(4), 227-233.

Nordness, P. D., Haverkost, A., & Volberding, A. (2011). An examination of hand-held computer-assisted instruction on subtraction skills for second grade students with learning and behavioral disabilities. *Journal of Special Education Technology,* 26(4), 15-24.

Obiakor, F. E., & Yawn, C. D. (2014). Reducing achievement gaps and increasing the school success of culturally and linguistically diverse students with special needs using the comprehensive support model. In C. M. Camille & S. D. Horsford (Eds.), Advancing equity and achievement in America's diverse schools (pp. 159-170). New York: Routledge.

Payne, L. D., Marks, L. J., & Bogan, B. L. (2007). Using curriculum-based assessment to address the academic and behavioral deficits of students with emotional and behavioral disorders. *Beyond Behavior,* 16(3), 3-6.

Pierce, M. E., & Fontaine, L. M. (2009). Designing vocabulary instruction in mathematics. *Reading Teacher, 63*(3), 239-243.

Ryan, J. B., Pierce, C. D., & Mooney, P. (2008). Evidence-based teaching strategies for students with EBD. *Beyond Behavior*, 17(3), 22-29.

Seifert, K., & Espin, C. (2012). Improving reading of science text for secondary students with learning disabilities: Effects of text reading, vocabulary learning, and combined approaches to instruction. *Learning Disability Quarterly,* 35(4), 236-247.

Sibold, C. (2011). Building English Language Learners' academic vocabulary strategies and tips. *Multicultural Education,* 18(2), 24-28.

Trout, A. L., Nordness, P. D., Pierce, C. D., & Epstein, M. H. (2003). Research on the academic status of children with emotional and behavioral disorders: A review of literature from 1961 to 2000. *Journal of Emotional and Behavioral Disorders,* 11, 198–210.

In: S.T.E.M. Education
Editor: Satasha L. Green

ISBN: 978-1-63117-806-1

Chapter 6

TEACHING STEM TO STUDENTS WITH AUTISM SPECTRUM DISORDERS

***Ernst Van Bergeijk*[*1], *Michelle Ranaldo*[2] *and Oren Shtayermman*[3]**
New York Institute of Technology, US

ABSTRACT

Early identification and intervention of students with Autism Spectrum Disorders (ASDs) has led to better outcomes for this population. In a previous generation students with Autism were not a part of general education, let alone preparing for college in the areas of science, technology, engineering, and mathematics (STEM). Now, teachers of STEM are faced with the unique challenges and rewards of teaching this population. Teachers of STEM content areas will need to work with a variety of professional support staff such as school social workers, psychologists, speech and language therapists, and occupational therapists to address some of these issues of students with ASD. A crucial role teachers of STEM can play is that of a mentor, helping guide the students through the complex maze of the social environment. In the mentoring role teachers can encourage the development of talents in STEM and advise regarding the readiness to transition to 4-year colleges and universities. With this population, developing their talent in Art should be a consideration. Teachers should think in terms of STEAM - Science, Technology, Engineering, Art, and Mathematics when working with students on the Autism spectrum.

INTRODUCTION

Both the popular press and scholarly researchers have documented an "explosion" in the incidence of cases of Autism and Autism Spectrum Disorders (ASD) within the past decade, labeling it as an "epidemic" (Cowley, Underwood, Murr, Springen, & Sennott, 2003; Ehlers & Gillberg, 1993; Bryson & Smith, 1998; Fombonne, 2003; Fording, 2003; Nash, 2002; Ozonoff, Rogers, & Pennington, 1991). Before the 1990s, the prevalence of Autism was

[*] evanberg@nyit.edu.

thought to be 1 in 10,000 (Brasic, 2001; Klin, Volkmar, & Sparrow, 2000) making this spectrum of disorders relatively rare. Epidemiological studies on ASD were first conducted in England during the mid 1960s' (Fombonne, 2005). The prevalence of ASD 60 years ago when it was first discovered was approximately 0.4 per 1,000 children (Prior, 2003). Studies conducted in the past decade have estimated the prevalence of ASD as ranging from 0.4 to almost 6 per 1,000 children, depending on the place, time, country, and the population studied (Prior, 2003). The current prevalence rates of ASD range from 0.5 to 6.7 per 1,000 among children ages 3 through 10 years (Bertrand, Boyle, Yeargin-Allsopp, Decoufle, Mars & Bove, 2001; Charman, 2002; Croen, Grether, Hoogstrate & Selvin, 2002; Fombonne, 2003; Wing, 1993). More specifically, prevalence of Autistic Disorder was found to be between 0.07 to 7.2 per 1,000 children.

Recent evidence suggests that ASD prevalence has been increasing over time, as evidenced by higher prevalence rates in successive birth cohorts. The increases were greatest for annual cohorts born between 1987 and 1992 (Newschaffer, Falb, & Gurney, 2005). The prevalence rate of ASD in children was as high as one in 166 in 2005 compared with the prevalence rate of one in 2,500 ten years ago (Center for Disease Control, 2005). More recently the Center for Disease Control (CDC) published the results of a telephone survey of parents and found the prevalence to be 1 out of 50 school aged children has a diagnosis of an ASD (Blumberg et al., 2013). The increase in the prevalence of ASD is partially due to changes in the definitions of each of the disorders on the Autism Spectrum and better assessment tools using standardized diagnostic measures (Fombonne, 2005). Only recent epidemiological studies on ASD include data on Pervasive Developmental Disorders and Asperger's syndrome. This could serve as an additional explanation to the increase in the prevalence of ASD (Fombonne, 2005).

What Are Autism Spectrum Disorders?

Autism Spectrum Disorders (ASD) are a group of life-long developmental disabilities caused by an abnormality of the brain (Rutter, 2005). Autism Spectrum Disorders are also known as neurodevelopmental disorders or neurobehavioral disorders (Bertrand, Boyle, Yeargin-Allsopp, Decoufle, Mars, & Bove, 2001; Croen, Grether, Hoogstrate & Selvin, 2002). Autism was first introduced to the DSM-III in 1980 as part of the Pervasive Developmental Disorders to convey that individuals with these conditions suffered from impairment in the development and multiple areas of functioning (Volkmar & Klin, 2005). A synonym for Pervasive Developmental Disorders is Autism Spectrum Disorder. Asperger's Syndrome (AS) is considered a severe and chronic developmental disorder and it is part of the disorders on the Autism spectrum (Klin, McPartland & Volkmar, 2005). What distinguishes Asperger Syndrome from other Autism Spectrum Disorders are the higher linguistic and cognitive capacities. There is still confusion over and controversy with the nosological status of AS as a separate condition from Autism and whether AS is a distinct diagnosis from Autism or a subtype of Autism (Lord & Corsello, 2005). The term is also used to describe individuals with Autism with higher verbal abilities, but who are socially disabled.

Autism Spectrum Disorders (ASD) are conceptualized as a continuum of syndromes ranging from Autistic Disorder to Pervasive Developmental Disorder Not Otherwise Specified (PDD-NOS). The continuum includes Rett's Disorder, Childhood Disintegrative

Disorder, Atypical Autism, and Asperger's Disorder (DSM-IV-TR, 2000). The American Psychiatric Association added Asperger's Syndrome to the *Diagnostic and Statistical Manual of Mental Disorders, 4th Edition (DSM-IV)* using the umbrella rubric of Pervasive Developmental Disorders (Barnhill, Tapscott Cook, Tebbenkamp, & Smith Myles, 2002), which is alternatively known as "Autism Spectrum Disorders" (Wing, 1991). Among the Autism community High Functioning Autism (HFA) is used to distinguish between individuals with classic Autism symptoms (i.e., those with little or no speech) and those with speech that had a late onset.

Asperger's Syndrome is one of the five categories of ASD. Most studies of comorbidity among individuals with ASD have used case reports or case series (Martin, Patzer & Volkmar, 2000). Asperger's syndrome (AS) is an ASD that is characterized by highly impaired social skills, difficulty relating to others, a lack of flexible imaginative play, and often a preoccupation with a finite and highly specific topic. AS, which typically presents less severe symptoms than classic Autism, has seen a dramatic increase in its prevalence within the last 10 years. The current diagnostic features of AS include a qualitative impairment in social relationships. The diagnosis of AS typically occurs much later than Autism.

Individuals with Autism are diagnosed in early childhood because of their profound inability to interact with others. The diagnosis of Autism can be made as early as 18 months and can be made reliably by 30 months of age (Gillberg, Nordin, & Ehlers, 1996). Children with AS are diagnosed on average between 6-10 years old, only after having years of difficulties learning and relating to peers (Gillberg, Nordin, & Ehlers, 1996). Children diagnosed with AS are socially isolated and this isolation cannot be explained by other personality features such as shyness or aggressiveness (Barnhill, Taspscott Cook, Tebbenkamp, & Smith Myles, 2002; McLaughlin-Cheng, 1998; Klin, Volkmar, & Sparrow, 2000). What limited social interaction they do engage in revolves around their own obsessive interests. Their impairment not only lies in a social skills realm, but also in speech. Although children with AS acquire speech, they have impairments in pragmatics and semantics of speech. They have difficulty providing a listener with relevant contextual information.

As of May 2013, the American Psychiatric Association (APA) revised the Diagnostic Manual of Mental Disorders known as the DSM 5. The APA eliminated the pervasive developmental disorders. According the APA, ASD is a new DSM-5 name that reflects a scientific consensus that four previously separate disorders are actually a single condition with different levels of symptom severity in two core domains. ASD now encompasses the previous DSM-IV Autistic Disorder (Autism), Asperger's Disorder, Childhood Disintegrative Disorder, and Pervasive Developmental Disorder not otherwise specified. ASD is characterized by (1) deficits in social communication and social interaction and (2) restricted repetitive behaviors, interests, and activities (RRBs). Because both components are required for diagnosis of ASD, social communication disorder is diagnosed if no RRBs are present (APA, 2013).

The impact of the change in diagnoses is currently unknown. Some researchers contend it will have little impact upon the number of children who will be diagnosed with an ASD (Huerta, Bishop, Duncan, Hus, & Lord, 2012). Others fear that many higher functioning individuals with Asperger Syndrome or PDD-NOS will no longer meet the diagnostic criteria, and as many as 23% of these individuals could be affected (Gibbs, Aldridge, Chandler, Witzlsperger, & Smith, 2012). What is also unknown is the impact the change in the DSM will have upon the educational system and how students qualify for special education services under the Individuals with Disabilities Education Act (IDEA). It may take many years for the professional community to change over to the new system of nomenclature and for the effects of the change to be felt.

Teachers of STEM are unlikely to come into contact with many students on the Autism Spectrum who are lower functioning. By lower functioning we mean students with full scale I.Q. scores below 70, which indicate an intellectual disability. It is unlikely that teachers of STEM will have students with classic Autistic Disorder, many of whom are non-verbal as well as having an Intellectual Disability. Likewise, teachers of STEM will probably not have students with Rett's, and Childhood Disintegrative Disorder, all of which are associated with lower I.Q.s scores and are lower functioning. Teachers of STEM are most likely to teach students with diagnoses such as Asperger Syndrome, Pervasive Developmental Disorder-Not Otherwise specified, or atypical Autism which is often colloquially referred to as High Functioning Autism (HFA).

Co-Morbid Disorders

With students on the Autism Spectrum who are higher functioning, the challenge will not be teaching the core material. These students are often keenly interested in the subject matter and can have an encyclopedic depth of knowledge about the subject. Many of them excel at rote knowledge and can memorize an astonishing amount of material. Rather, for the teacher the challenge will be helping the student navigate the social aspects of the classroom and learning and cope with their possibly co-morbid psychiatric issues. Teachers will need the support and guidance of school based mental health professionals such as school psychologists, social workers, and counselors. It is important that teachers receive training in the signs and symptoms as well as the management of these co-morbid disorders.

Depressive and anxiety disorders are the most prevalent psychiatric disorders reported among individuals with ASD (Lainhart, 1999). Some reports indicate that psychotic conditions and conduct disorder may also be present in individuals with Autism (Klin, McPartland, &Volkmar, 2005). It is difficult to diagnose depression in individuals with Autism because assessment of depression relies mainly on communication skills. Volkmar and Klin (2005) noted that many individuals with Autism are also diagnosed with mental retardation. Edelson (2006) stated that the average prevalence of mental retardation among individuals diagnosed with Autism is 79%. Estimates of the prevalence of depression in individuals with ASD range widely from 5% to 82% (Barnhill, 2001; Ghaziuddin, Weidmer-Mikhail, & Ghaziuddin, 1998; Ghaziuddin, Alessi, & Greden, 1995; Green, Gilchrist, Burton, & Cox, 2000; Kim, Szatmari, Bryson, Streiner, & Wilson, 2000; Klin, Pauls, Schultz, & Volkmar, 2005).

Indeed, individuals diagnosed with ASD are often diagnosed with additional symptoms or disorders not part of the ASD diagnosis (Ghaziuddin, Weidmer-Mikhal & Ghaziuddin, 1998; Gillberg & Billstedt, 2000; Tantam, 2000). Psychiatric disorders that are comorbid with Autism Spectrum Disorders have not been adequately studied (Bryson & Smith, 1998; Tantam, 2003). Most of the studies that have investigated comorbidity in Autism have been based on case reports or case series (Ghaziuddin, Weidmer-Mikhal & Ghaziuddin, 1998; Martin, Patzer & Volkmar, 2000). The studies are methodologically weak with mostly small nonprobability samples. Anxiety disorders, such as Generalized Anxiety Disorder and

Separation Anxiety, were reported in 7% to 50% of individuals with ASD (Lainhart, 1999). In review of studies about ASD and comorbid disorders, the range of prevalence of anxiety disorders was 7.3% to 35% (Gadow, DeVincent, Pomeroy & Azizian, 2004; Green, Gilchrist, Burton & Cox, 2000; Kim, Szatmari, Bryson, Streiner & Wilson, 2000; Klin, Pauls, Schultz & Volkmar, 2005; Tonge, Brerton, Gray & Einfeld, 1999). All of these samples were drawn from specialty clinic and treatment centers.

Approximately 250,000 adolescents attempt suicide each year, and 8% to 10% of all children in the United States attempt suicide at some point during their childhood (Horowitz et al., 2001). Worldwide, suicide is ranked as the thirteenth most common cause of death among adolescents, following accidental injuries, homicide, AIDS, and war (Blum, 2005). In the United States, suicide is the third leading cause of death among adolescents (Center for Disease Control, 2004). Between the years 1980 and 1992, suicide was the third leading cause of death for those who were 13 to 19 years old, and the third leading cause of death for those who were 15 to 24 years old in the U.S (Barrios, Everett, Simon & Brener, 2000; Bloch, 1999; Rosewater & Burr, 1998). When data on suicide in the general population are presented or discussed, there is usually a concern regarding their validity (Evans, Hawton, Rodham & Deeks, 2005). The main difficulties in obtaining accurate data are: (a) inconsistent definitions of suicidal behavior, (b) lack of a system of classification for suicidal acts, and (c) the type of sources reporting on the data. In previous research, data have been hospital admissions records, epidemiological surveys and psychiatric samples (Evans, Hawton, Rodham & Deeks, 2005). In some countries, suicide is hidden due to cultural influences, with actual rates of suicide being higher than reported (Rudmin, Ferrada-Noli & Skolbekken, 2003). According to Shtayermman (2006), adolescents and young adults diagnosed with AS, have levels of suicidal ideation comparable to those of neurotypical adolescents or the young adults. Recent research found that, in fact, suicidal ideation and attempts are 28 times greater among children with ASDs than the neurotypical population. However, children with ASDs are still 3 times less likely to exhibit suicidal ideation and attempts than student with depression alone (Dickerson Mayes, et al., 2013).

Tourette's Syndrome (TS) in the general population is thought to be approximately 1 per 169 people (Ringman & Jankovich, 2001). The co-occurrence of both Tourette's and AS would be 5 per 100 million if the two disorders were to occur randomly (Baron-Cohen, Mortimore & Moriarty, 1999). In actuality, the prevalence of Tourette's among the AS population is significantly higher. Volkmar, Klin, Schultz, Pauls, and Cohen (1996) recognized 2% of their AS study participants as also having Tourette's (as cited in Klin, Sparrow, Marans, Carter, & Volkmar, 2000). This is almost four times higher than the prevalence in the general population. Ehlers and Gillberg (1993) identified 20% of school-aged children with co-morbid AS and Tourette's in their Swedish general population study (as cited in Gillberg & Billstedt, 2000). A second Swedish population study only revealed 10% of the children who had Tourette's also had a diagnosis of AS (Kadesjö & Gilberg, 2000). Baron-Cohen et al. (1999) contends that the co-morbidity of Tourette's Syndrome and AS far exceeds the rate of TS and Autism. They estimate that 50% of individuals with AS also have TS, whereas only between 2.6-20.3% of individuals with Autism have Tourette's.

The unpredictability of the social environment can cause some individuals with Autism to develop anxiety disorders. Compulsive rituals among individuals with Autism lead psychiatrists to think that this may be a display of obsessive-compulsive disorder (OCD) (Tantam, 2000) and they may in fact have co-morbid OCD. Klin and Volkmar (1997) identified 19% of their sample of 99 AS patients as having OCD (as cited in Martin, Patzer, & Volkmar, 2000). Social phobia is a sub category of anxiety disorders where the individual displays a persistent fear of social interactions or public performance. The resultant behavior can be a panic attack when thinking about or confronted with a social situation, with which they are unfamiliar. Tantam (2000) noted that adolescents with social phobia may, in fact, have undiagnosed cases of AS, especially when there is an absence of a complete developmental history.

Children with Autism are more likely to be diagnosed with Attention Deficit Hyperactivity Disorder (ADHD) as a secondary diagnosis or even as an initial diagnosis than either teens or adults. This is not surprising given the fact that AS and ADHD share similar features and ADHD is a diagnosis of early childhood. Klin and Volkmar's (1997) study of 99 individuals with AS found that 28% of their sample exhibited ADHD (as cited in Martin, Patzer, & Volkmar, 2000). A survey of the Online Asperger Syndrome Information and Support Website (OASIS) found that ". . . 65% [of parents] indicated that their child had been initially described as one or another type of ADHD, while 44% stated that their child had a dual diagnosis of AS and a form of ADHD" (Romanowski-Bashe & Kirby, 2001, p. 69). Another survey of 514 parents of children with AS found that 42.8% had ADHD. An additional 30% of the AS children had a diagnosis of Attention Deficit Disorder (ADD) (Romanowski-Bashe & Kirby, 2001).

It has been suggested that there is considerable overlap between the clinical manifestations of AS and a Nonverbal Learning Disability (Rourke & Tsatsanis, 2000). The overlap is seen in the limited appreciation of humor, limited ability to understand or use the rules governing social behavior, and poor apprehension and application of nonverbal aspects of communication (Rourke & Tsatsanis, 2000).

Individuals with a Nonverbal Learning Disability are especially at risk for development of an internalized socio-emotional pathology, including withdrawal, anxiety, and depression (Bender et al., 1999; Kowalchuck & King, 1989). It was also noted by Bender et al. (1999) that individuals with a Nonverbal Learning Disability are also at higher risk for depression and suicide than individuals with verbal disability (Bender et al., 1999).

Evidence suggests that the cognitive profiles of individuals with AS and individuals with a Nonverbal Learning Disability are similar (Barnhill, Hagiwara, Myles & Simpson, 2000), making the diagnosis of AS more complex (Rourke & Tsatsanis, 2000). Some of the similarities between these syndromes are the marked deficits in nonverbal problem solving and deficiencies in age-appropriate sensitivity to humor (Rourke & Tsatsanis 2000). A Nonverbal Learning Disability is typically associated with the right hemisphere dysfunction, which affects the ability to develop effective adaptation to environmental limitations and, in turn, places constraints on coping strategies. The clinical features of a Nonverbal Learning Disability (NLD) and AS have similar patterns of behaviors and adaptive functioning as well. However, there are noticeable differences between Nonverbal Learning Disability and AS in terms of the manifestation of social disabilities (Rourke & Tsatsanis, 2000) and unlike individuals with Nonverbal Learning Disability, individuals diagnosed with AS also exhibit restricted, repetitive and unusual patterns of behaviors and interest (Little, 2002).

IMPORTANCE OF STEM INSTRUCTION FOR LEARNERS WITH AUTISM

The U.S. Department of Education's 1983 national report *Nation at Risk* stated that students in the United States were falling way below the rest of the world in preparedness for the changing workforce. Declining student numbers in science, mathematics and engineering majors, low SAT scores and literacy rates added to the rapidly declining competition within the world's global economy, ("A nation accountable:," 2008). It was recommended that Education initiatives provide for reform in those areas that would help students attain the necessary skills and knowledge to compete equally with their peers in a global economy (Martin, Stumbo, Martin, Collins, Hedrick, Nordstrom, & Peterson, 2011). Our nation ranks 27th in the world market and the rankings are even lower for those with disabilities.

In 2006, the National Academies initiative for advancing STEM instruction was also considered by federal policy makers and recommendations were made to continue to reform education practices and provide focused instruction. In addition, many industry leaders promote education reform in STEM areas in order to recruit innovative and creative thinking workers and consumers (Israel, Maynard, Williamson, 2013).

PEDAGOGICAL APPROACHES TO STEM FOR THIS POPULATION

Understanding the fundamental proficiencies required to participate in STEM activities is key to providing appropriate instruction. Instructors at all grades need to have a clear understanding of the four fields and how they are intertwined (Basham & Marino, 2013). Applying Universal Design for Learning to instructional strategies can assist in successful implementation of STEM instruction for learners with Autism. STEM requires students to learn problem solving strategies and complex content. Basham et al. suggests that K-12 instruction focusing on an engineering design. In other words focus on developing the appropriate knowledge and strategies in order to solve problems. The engineering design process requires students to develop inquiring minds and build on that knowledge in order to find solutions to various problems.

It is thought that many learners on the Autism spectrum have a predilection towards science and mathematics because they are rule based and can recite and memorize lists as well as calculate easily in their heads (Moon, Todd, Morton, & Ivey, 2012). However, struggles with STEM content usually presents itself in the form of having the ability to develop higher-order thinking skills, find relationships between subject areas, collaborative work and problem solving (Basham & Marino, 2013). In order to develop higher order thinking skills students must be able to work with prior knowledge, tackle problems and seek solutions. However, students with ASDs are capable of higher level work and can excel even at a college level. The student will need assistance in areas of executive functioning or meta-organizational skills. Large complex assignments may need to be broken down into smaller more manageable tasks. Timers and reminders may need to be programmed into a smart phone or personal digital assistant. Binders may need to be color coded to differentiate syllabi, class notes, research, draft assignments etc. (Van Bergeijk, Klin, & Volkmar, 2008).

Temple Grandin is probably the most famous person on the Autism spectrum currently living. She has a Ph.D. in animal sciences and has designed approximately 40% of the abattoirs in the U.S. Not only is she a prolific writer in the animal sciences, but she is also a leading authority on Autism. Grandin contends that there are three different types of brains of individuals on the Autism spectrum: (1) visual thinkers; (2) music and higher math brains; and (3) verbal lists and translator brains (Grandin & Duffy, 2004). Her own visual ability enabled her to re-design slaughter houses in the U.S. to be more humane and less frightening to livestock. This ability and how she learned are the title and subject of her book: *Thinking in Pictures* (1996). Grandin and Duffy (2004) recommended having the student with Autism develop natural talents and to use those talents to incorporate the academic material to be mastered. For example, for those students who draw well could either draw free hand or through Computer Aided Design (CAD) aspects of an engineering project. Teachers of STEM ought to consider encouraging development of the talents of student with ASDs in the area of Art. Some of these students like Temple Grandin have extraordinary visual skills and artistic talent. As educators in the sciences we need to think in terms of STEAM, adding art to our assessment and instruction of students with ASDs.

Grandin and Duffy's (2004) book provides examples of careers best suited for the different types of autistic brain as well as advice on how to gain entry into those careers and how to handle job interviews including small talk. Students on the spectrum seem to relax and perform well when they are able to discuss projects from a portfolio as an ice breaker activity when working in small groups on a project (Grandin & Duffy, 2004). Traditional ice breakers like talking about oneself (i.e. small talk) are anxiety provoking for individuals on the spectrum because they do not understand the social rules about disclosure, not dominating a conversation and their inability to read non-verbal cues.

In working with individuals on the Autism spectrum it is critical that they be given practical opportunities to apply what they have learned in real world settings. Internships in laboratories, engineering firms, or working with professors conducting research are invaluable experiences for these students. Mentoring is critical to successfully learning not only the STEM material, but how to negotiate the social world of small group projects, college, and the world of work (Grandin & Duffy, 2004). Students on the spectrum need coaching on how to handle the politics of the workplace and how to make small talk with their peers. Without mentoring, internships, and paid work experiences during high school, it is extremely difficult for students on the Autism spectrum to secure post-secondary school employment in STEM or any other field. In fact, key predictors of post-secondary employment for students on the Autism spectrum is having a job for pay during high school, having job training as a part of high school, and having the school provide vocational counseling including contacting employers (Chiang , Cheung, Li, & Tsai, 2012).

Mentoring must be explicit in terms of the steps necessary to make the transition from a secondary school environment to the complex environment of a college campus. In high school all the classes are contained generally in one building. External cues such as bells signal to students when they are to move on to their next class. The day is high structured. The student's schedule is full and contains very little free time. Unstructured free time can be the downfall of students with ASDs. Students on the Autism spectrum need to be taught how to put in place reminders to go to class, start a research project, and generally structure their free time. Mentors can play a critical role in this process. Likewise, mentors may need to coach students on independent living skills that are critical to successfully transitioning to living independently. These skills are not commonly taught at high schools. Higher functioning students on the Autism spectrum often fool educators because their vast academic knowledge particularly in areas related to STEM. What educators do not realize is that these students will need help learning how to do laundry, organize their dorm rooms, travel safely using mass transit (a high percentage of students on the Autism spectrum do not know how to drive a car), wake up on their own using an alarm clock or smart phone, and learning how to budget money. Many of these skills seem like common sense, but they must be explicitly taught to this population. Finally, mentors can pay a critical role by providing honest feedback and assessment of student's readiness to transition to living away from home at a 4-year college. Many students on the spectrum are not ready to handle the social and independent living skills aspects of going away to a 4-year college. Mentors can recommend 2-year community colleges to the students with ASDs who are not yet ready to live away from home. Attending a community college offers a number of advantages over 4-year colleges for students on the spectrum including significantly lower costs, the proximity of familial support, the familiarity of their home and community environment, and the practical, career oriented approach of associate degrees. Some students on the spectrum do not understand the concept of liberal arts degrees and refuse to take courses such as humanities or English since it does not appeal to their interests in computers, for example. There is a more direct connection between the course content of associate degrees and a chosen career than a 4-year liberal arts degree. Counseling a student with an ASD to pursue an associate degree in STEM is empirically supported. According to Wei, Christiano, Yu, Blackorby, Shattuck, and Newman (2013), students with an ASD who pursue an associate degree in STEM are more likely to complete the associate degree than students on the spectrum who pursue other majors. Furthermore, students on the spectrum who complete an associate degree in STEM are also more likely to complete a bachelor degree than students with ASDs who pursue other majors Wei, Christiano, Yu, Blackorby, Shattuck, and Newman, (2013) found:

> ...most college students with an ASD enrolled in a 2-year community college at some point in the postsecondary careers (81 %). Those in STEM fields were more likely to persist in a 2-year community college and were twice as likely to transfer from a 2-year community college to a 4-year university than their peers in the non-STEM fields.

Incorporating STEM Instruction in K-12 Grades

In order for there to be successful integration of STEM content teachers must have a comfort level with the subject matter, time to design curriculum and have a working understanding of universal design for developing instructional materials (Basham & Marino, 2013). Along with a very clear understanding of a student with Autism's educational needs there are many ways to incorporate customized instruction. According to the research findings conducted by the Center for Assistive Technology and Environmental access students on the spectrum will benefit from:

- Graphic organizers with which to visualize connections and relationships
- Time management techniques for breaking down complex tasks into manageable tasks
- Charts and graphics to illustrate concepts
- Vocabulary lists

- Clear concise instructions
- Study guides
- Discussions and directions for completing research techniques
- Distraction-free environments
- Access to technology for creating notes, drawings storing and accessing information
- Course management through web-based instructional software
- Online bulletin boards
- Social media (Moon, N. W., Todd, R. L., Morton, D. L., & Ivey, E. , 2012).

K-12 Strategies and Resources

Students in grades as early as preschool have opportunities to work with STEM concepts. Curriculum that provides experiences and practice with vocabulary development, collaboration activities, making connections, inquiry and building scientific relationships helps to develop those skills necessary for future learning (McGough & Nyberg, 2013).

Table 1. Additional Curriculum Resources

Resource or Organization	Curriculum	Website/Publication
National Research Council (NRC)	A Framework for K-12 Science education: Practices, crosscutting concepts and core ideas.	ISBN-10: 0-309-21742-3
National Science Foundation	Diverse educational resources in science, engineering and technology	www.nsf.gov/discoveries/
National Center on Accessible Instructional Materials	Resources for Instruction and Collaboration	aim.cast.org
Edutopia.org	Creative and evidence based learning strategies in STEM including assessment and problem based learning	www.edutopia.org
edWeb	International social and learning community of professionals working with students on the Autism spectrum	www.edweb.net

It is recommended that curriculum be infused with technology. Students as early as preschool already have experience with a wide variety of technological devices. Integration of a technology rich curriculum provides resources for many students that are not otherwise accessible. Digitized textbooks, videos, interactive simulations and applications that support executive functioning skills give teachers a broader foundation of materials that support instruction and learning (Basham & Gardner, 2013). With the help of tablets and especially the iPad there is a multitude of applications that introduce STEM concepts and experiences with students as early as preschool age. Autism Speaks an internationally recognized organization recommends applications and resources that support communication, organization, social skills, math skills, and creative arts. Additional recommendations and resources include: (a) Minecraft simulation incorporates engineering design practices, (b) Community based projects – solutions to real-world problems such as recycling and self-sustaining ecosystems, (c) Robotics, (d) Game based learning, (e) Digital resources such as videos and podcasts, (f) Constructs with real-world materials, and (g) Modeling.

Although the focus has been on STEM experts are now working on incorporating art into the equation. Art provides the same creativity and design characteristics present in the engineering process (Bequette & Bequette, 2012). Students on the Autism spectrum often gravitate towards an artistic form of expression. Including the creative methods and process of design, function and media students have the opportunity to exercise their imagination while applying the principles of engineering design (Wynn & Harris, 2012).

Differentiating Instruction for Learners with ASD

Reinforcing the practice of Universal Design with instructional tools and strategies Teachers need to be cognizant of their learner's weaknesses and strengths. Many learners on the spectrum may not always self-identify in inclusive classrooms especially in middle school and secondary classrooms. While students may have the intellectual capacity for higher order thinking skills they often do not have the ability necessary to participate in group based work projects involving inductive reasoning (Moon, Todd, Morton & Ivey, 2012).

The Alliance for students with disabilities in STEM is a leader in promoting universal design instruction (Access STEM, 2013). For instruction to be effective it should be of equitable use. All who have access to instruction should have the capability or accommodations in place to learn. The instructional design should take into consideration a side variety of abilities and preferences. Whether the instruction is available to read or some accommodation is provided to access text to speech. The design should be simple and intuitive. Clear instructions are provided and knowledge is scaffolded with prior knowledge. The information is perceptible to the learner. Often students on the spectrum have difficulties with various sensory conditions. Temperature, background noise and distractions impede the acquisition of knowledge. Provide feedback and tolerance for error. The design should include directives or assistance in finding the correct answer or procedures. The design should be able to be used efficiently and comfortably, and with a minimum amount of exertion (Access STEM, 2013).

Conclusion

Careers requiring skills in STEM areas are often underrepresented by those on the Autism spectrum not because there is not capability, but often due to the lack of understanding the supports needed within the school system. Early identification and intervention with empirically based techniques has led to better outcomes for children on the Autism spectrum. A generation ago, many of these same children would have been relegated to psychiatric hospitals with no hope of obtaining an education and earning a college degree. Now, these very children are able to participate in public education and go on to earn degrees in STEM where many of them excel. Their different world view will lead to innovations in science, technology, engineering, and mathematics. They add to the neurodiversity of our society's collective thought processes. Some of these very children who would have been discarded a generation ago will contribute to the next generation's breakthroughs. As teachers of STEM to this talented, but at times challenging population, the trick will not be how to teach the core material. Rather, it will be teaching students on the Autism spectrum how to deal with the non-academic aspects of the curriculum including social rules of working in groups and being a co-worker, living independently, and how to successfully enter and stay in the world of work. The role of mentor or life coach is a critical role that teachers of STEM can provide the students on the Autism spectrum. It is in this role that teachers will have their greatest impact upon this population.

REFERENCES

Access STEM. (2013). Retrieved from http://www.washington.edu/doit/Stem/ld.html

American Psychiatric Association. (2000). *DSM-IV-TR.* Washington, D.C.: Author.

American Psychiatric Association. (2013). Highlights of changes from the DSM-IV TR to DSM 5. Retrieved from http://www.dsm5.org/Documents/changes%20from%20dsm-iv-tr%20to%20dsm-5.pd

Aronin, S., & Floyd, K. K. (2013). Using an iPad in inclusive preschool classrooms to introduce STEM concepts. *Teaching Exceptional Children, 45*(4), 34-39.

Barnhill, G. P. (2001). Social attribution and depression in adolescents with asperger syndrome. *Focus on Autism and other Developmental Disabilities, 16*(1), 46-54.

Barnhill, G., Hagiwara, T., Myles, B. S., & Simpson, R. L. (2000). Asperger syndrome: *Focus on Autism and other Developmental Disabilities, 15*(3), 146-153.

Barnhill, G., Tapscott-Cook, K., Tebbenkamp, K., & Smith Myles, B. (2002). The effectiveness of social skills targeting nonverbal communication for adolescents with Asperger syndrome and related pervasive development delays. *Focus on Autism and Other Developmental Disabilities, 17,* 112–118.

Baron-Cohen, S., & Wheelwright, S. (1999). Obsession in children with Autism or asperger syndrome. *British Journal of Psychiatry, 175,* 484-490.

Baron-Cohen, S., Mortimore, C., Moriarty, J., Izaguirre, J., & Robertson, M. (1999). The Prevalence of Gilles de la Tourette's Syndrome in Children and Adolescents with Autism. *Journal of Child Psychology and Psychiatry*, 40, pp 213-218.

Barrios, L. C., Everett, S. A., Simon, T. R., & Berner, N. D. (2000). Suicide ideation among US college students. *Journal of American College Health, 48,* 229-233.

Basham, J. D., & Marino, M. T. (2013). Understanding STEM education and supporting students through universal design for learning. *Teaching Exceptional Children, 45*(4), 8-15.

Basham, J. D., Israel, M., & Maynard, K. (2010). An ecological model of STEM education: Operationalizing STEM FOR ALL. *Journal of Special Education Technology, 25*(3), 9-19.

Bashe, P. R., & Kirby, B. L. (2001). *The OASIS Guide to Asperger syndrome: Advice, Support, Insights, and Inspiration.* Crown Pub

Bender, W. N., Rosenkrans, C. B., & Crane, M.K. (1999). Stress, depression, and suicide among students with learning disabilities. *Learning Disability Quarterly, 22*(2), 143-156.

Bequette, J., & Bequette, M. (2012, March). *Art Education*, 40-47.

Berkeihiser, M., & Ray, D. (2013). Bringing STEM to Life. *Technology & Engineering Teacher, 72*(5), 21-24.

Bertrand, J., Boyle, C., Yeargin-Allsopp, M., Decoufle, P., Mars, A., & Bove, F. (2001). Prevalence of Autism in a United States Population. *Pediatrics, 108*(5), 1155-1162.

Bloch, D. S. (1999). Adolescent suicide as a public health threat. *Journal of Child and Adolescent Psychiatric Nursing, 13*(1), 26-38.

Blum, R. (2005). *Adolescent health.* Retrieved from http://usinfo.state.gov/journals /itgic/0105/ijge/blum.htm.

Blumberg, S. J., Bramlett, M.D., Kogan, M.D., Schieve, L.A., Jones, J.R., &. Lu, M.C. (2013). Changes in Prevalence of Parent-reported Autism Spectrum Disorder in School-aged U.S. Children: 2007 to 2011–2012. National Health Report Statistics, 65: 1-7. Retrieved from http://www.cdc.gov/nchs/data/nhsr/nhsr065.pdf

Bryson, S. E., & Smith, I. M. (1998). Epidemiology of Autism: prevalence, associated characteristics, and implications for research and service delivery. *Mental Retardation Developmental Disabilities Research Review. 4*, 97–103.

Charman, T. (2002). The prevalence of Autism spectrum disorders. *European Child and Adolescent Psychiatry, 11*(6), 249-256.

Chiang, H. M., Cheung, Y. K., Li, H., & Tsai' L. Y. (2012). Factors Associated with Participation In Employment for High School Leavers with Autism. *Journal of Autism and Developmental Disorders* DOI: 10.1007/s10803-012-1734-2.

Cowley, G., Underwood, A., Murr, A., Springen, K., & Sennott, S. (2003). Girls, boys and autism. *Newsweek. 142,* (10).

Croen, L. A., Grether, J. K., Hoogstrate, J., & Selvin, S. (2002). The changing prevalence of Autism in California. *Journal of Autism and Developmental Disorders, 32*(3), 207-215.

DeJarnette, N. K. (2012). America's children: Providing early exposure to stem (science, technology, engineering and math) initiatives. *Education, 133*(1), 77-84.

Dickerson Mayes, S., Gorman, A.A., Hillwig-Garcia, J., & Syed, E. (2013). Suicidal ideation and attempts in children with Autism spectrum disorders. *Research in Autism Spectrum Disorders, 7*(1): 109 -119.

Edelson, M. G. (2006). Are the majority of children with Autism mentally retarded? A systematic evaluation of the data. *Focus on Autism and Other Developmental Disabilities, 21*(2), 66-83.

Ehlers, S., & Gillberg, C. (1993) The epidemiology of Asperger syndrome. A total population study. *Journal of Child Psychology and Psychiatry, 34, 1327–1350.*

Evans, E., Hawton, K., Rodham, K., & Deeks, J. (2005). The prevalence of suicidal phenomena in adolescents. *Suicide and Life-Threatening Behavior, 35*(3), 239-250.

Fombonne, E. (2003). The prevalence of Autism. *Journal of American Medical Association, 289*(1), 87-89.

Fombonne, E. (2005). Epidemiological studies of pervasive developmental disorders. In F. Volkmar, R. P., Klin, A., & Cohen, D. (Eds.), *Handbook of Autism and pervasive developmental disorders* (pp. 42-69). Hoboken, NJ: John Wiley and Sons, Inc.

Gadow, K. D., DeVincent, C. J., Pomery, J., & Azizian, A. (2004). Psychiatric symptoms in preschool children with PDD and clinic and comparison samples. *Journal of Autism and Developmental Disorders, 34*(4), 379-393.

Ghaziuddin, M., Alessi, N., & Greden, J. F. (1995). Life events and depression in children with pervasive developmental disorders. *Journal of Autism and Developmental Disorders, 25*(5), 495-502.

Ghaziuddin, M., Weidmer-Mikhail, E., & Ghaziuddin, N. (1998). Comorbidity of asperger syndrome. *Journal of Intellectual Disability Research, 42*(4), 279-283.

Gibbs, V., Aldridge, F., Chandler, F., Witzlsperger, E., & Smith, K. (2012). Brief report: an exploratory study comparing diagnostic outcomes for *Autism* spectrum disorders under *DSM*-IV-TR with the proposed *DSM*-5 revision. *Journal of Autism and Developmental Disorders, 42*(8), 1750-6.

Gillberg, C., & Billstedt, E. (2000). Autism and Asperger Syndrome. *Acta Psychiatrica Scandinavica, 102,* 321-330.

Gillberg, C., Nordin, V., & Ehlers, S. (1996). Early detection of Autism. Diagnostic instruments for clinicians. *Eur Child Adolesc Psychiatry, 5*(2), 67–74.

Grandin, T. (1996). *Thinking in pictures: And other reports from my life in Autism.* New York: Vintage Books.

Grandin, T., & Duffy, K. (2004). Developing Talents: Careers for individuals with Asperger syndrome and high functioning Autism. Shawnee Mission, KS: Autism Asperger Publishing Company.

Green, J., Gilchrist, A., Burton, D., & Cox, A. (2000). Social and psychiatric functioning in adolescents with asperger syndrome compared with conduct disorder. *Journal of Autism and Developmental Disorders, 30*(4), 279-293.

Horowitz, L. M., Wang, P. S., Koocher, G. P., Burr, B. H., Smith, M. F., & Klavon, S. (2001). Detecting suicide risk in a pediatric emergency department. *Pediatrics, 107*(5), 1133-1138.

Huerta, M., Bishop, S.L., Duncan, A., Hus, V., and Lord, C. (2012). Application of *DSM*-5 criteria for *Autism* spectrum disorder to three samples of children with *DSM-IV* diagnoses of pervasive developmental disorders. *The American Journal of Psychiatry , 169* (10): 1056-64.

Israel, M., Maynard, K., & Williamson, P. (2013). Promoting literacy-embedded, authentic STEM instruction for students with disabilities and other struggling learners. *Teaching Exceptional Children, 45*(4), 18-25.

K-12 science, technology, engineering, and math (STEM) education for America's future. (2011). *Tech Directions, 70*(6), 33-36.

Kadesjö, B., & Gillberg C. (2001). The comorbidity of ADHD in the general population of Swedish school-age children. *Journal Child Psychology and Psychiatry, 42, 87–92.*

Kennedy, M. J., & Wexler, J. (2013). Helping students succeed within secondary-level STEM content. *Teaching Exceptional Children, 45*(4), 26-33.

Kim, J. A., Szatmari, P., Bryson, S. E., Streiner, D. L., & Wilson, F. J. (2000). The prevalence of anxiety and mood problem among children with Autism and asperger syndrome. *Autism, 4*(2), 117-132.

Klin, A., Carter, A. S., Volkmar, F. R., Cohen, D. J., Marans, W. D., & Sparrow, S. S. (1997). Assessment issues in children with Autism. In D.J. Cohen and F.R. Volkmar (Eds.), *Handbook of Autism and Pervasive Developmental Disorders (2nd ed.).* (pp. 411-418). New York: John Wiley and Sons, Inc.

Klin, A., McPartland, J., & Volkmar, F. R. (2005). Asperger syndrome. In F. Volkmar, R. Paul, Klin Ami & D. Cohen (Eds.), *Handbook of Autism and pervasive developmental disorders* (pp. 88-125). Hoboken, NJ: John Wiley and Sons, Inc.

Klin, A., Pauls, D., Schultz, R., & Volkmar, F. (2005). Three diagnostic approaches to asperger syndrome. *Journal of Autism and Developmental Disorders, 35*(2), 221-234.

Klin, A., Volkmar, F. R., & Sparrow, S. S. (2000). *Asperger Syndrome.* New York: The Guilford Press.

Kowalchuk, B., & King, J. D. (1989), Adult suicide versus coping with nonverbal learning disorder, *Journal of Learning Disabilities, 22*(3), 177-178.

Kuchment, A. (2013, February 1, 2013). Students with Autism gravitate toward STEM majors. *Scientific American, 3*, 1-2.

Lainhart, J. E. (1999). Psychiatric problems in individuals with Autism, their parents and siblings. *International Review of Psychiatry, 11*(4), 278-299.

Leddy, M. (2010). Technology to advance high school and undergraduate students with disabilities in science, technology, engineering, and mathematics. *Journal of Special Education Technology, 25*(3), 3-8.

Little, L. (2002). Middle-class mothers' perceptions of peer and sibling victimization among children with Asperger's Syndrome and Nonverbal Learning Disability. *Issues in Comprehensive Pediatric Nursing, 25,* 43-57.

Lord, C., & Corsello, C. (2005). Diagnostic instrument in autistic spectrum disorders. In F. Volkmar, P. Klin & D. Cohen (Eds.), *Handbook of Autism and pervasive developmental disorders* (pp. 730-771). New York: Wiley and Sons.

Martin, A., Patzer, D. K., & Volkmar, F. R. (2000). Psychopharmacological treatment of higher-functioning pervasive developmental disorders. In A. Klin, F. R. Volkmar & S. S. Sparrow (Eds.), *Asperger Syndrome* (pp. 210-228). New York: The Guilford Press.

Martin, J., Stumbo, N., Martin, L., Collins, K., Hedrick, B., Nordstrom, D., & Peterson, M. (2011). Recruitment of students with disabilities: Exploration of science, technology, engineering and mathematics. *Journal of Postsecondary Education and Disability, 24*(4), 285-299.

McGough, J., & Nyberg, L. (2013). Strong stems need strong sprouts. *Science and Children,* 27-33.

McLaughlin-Cheng, E. (1998). Asperger syndrome and Autism: A literature review and meta-analysis. *Focus on Autism and Other Developmental Disabilities, 13*, 234-245.

Moon, N. W., Todd, R. L., Morton, D. L., & Ivey, E. (2012). *Accommodating Students with disabilities in science, Technology, engineering and Mathematics through university education (STEM) findings from research and practice for middle grades.* Atlanta, Georgia: SciTrain: Science and Math for All.

Moorehead, T., & Grillo, K. (2013). Celebrating the reality of inclusive STEM education. *Teaching Exceptional Children, 45*(4), 50-57.

Nadelson, L. S., Callahan, J., Pyke, P., Hay, A., Dance, M., & Pfiester, J. (2013). Teacher STEM perception and preparation: Inquiry-based STEM professional development for elementary teachers. *Journal of Educational Research, 106*(2), 157-168.

Newschaffer, C. J., Falb, M. D., & Gurney, J. G. (2005). National Autism prevalence trends from United States Special Education Data. *Pediatrics, 115*(3), 277-282.

O'Neil, T., Yamagata, L., Yamagata, J., & Togioka, S. (2012). Teaching STEM means teacher learning. *Kappan, 94*, 36-40.

Ozonoff, S., Pennington, B.F., & Rogers, S.J. (1991). Executive function deficits in high-functioning autistic individuals: Relationship to theory of mind. *Journal of Child Psychology and Psychiatry, 32*(7), 1081-1105.

PREPARE AND INSPIRE K-12 science, technology, engineering, and math (STEM) education for America's future. (2010). *Education Digest, 76*(4), 42-46.

Prior, M. (2003). Is there an increase in the prevalence of Autism spectrum disorders? *Journal of Pediatric. Child Health, 39,* 81-82.

Ringman J. M., & Jankovic J. (2000). Occurrence of tics in Asperger's syndrome and autistic disorder. *Journal of Child Neurol, 15*(6), 394–400.

Rockland, R., Bloom, D. S., Carpinelli, J., Burr-Alexander, L., Hirsch, L. S., & Kimmel, H. (2010). Advancing the "E" in K-12 STEM education. *Journal of Technology Studies, 36*(1), 53-64.

Rosewater, K. M., & Burr, B. H. (1998). Epidemiology, risk factors, intervention, and prevention of adolescent suicide. *Adolescent Medicine, 10*(4), 338-343.

Rourke, B. P., & Tsatsanis, K. D. (2000). Nonverbal learning disabilities and asperger syndrome. In A. Klin, F. R. Volkmar & S. S. Sparrow (Eds.), *Asperger Syndrome* (pp. 231-253). New York: The Guilford Press.

Rudmin, F. W., Ferrada-Noli, M., & Skolbekken, J. A. (2003). Questions of culture, age and gender in epidemiology of suicide. *Scandinavian Journal of Psychology, 44,* 373-381.

Rutter, M. (2005). Genetic influences and Autism. In F. Volkmar, R. Paul, Klin Ami & D. Cohen (Eds.), *Handbook of Autism and pervasive developmental disorders* (pp. 425-452). Hoboken, NJ: John Wiley and Sons, Inc.

Shah, N. (2012). Students with Autism. *Education Week, 32*(13), 5-5.

Shtayermman, O. "An exploratory study of suicidal ideation and comorbid disorders in adolescents and young adults with Asperger's syndrome" (January 1, 2006). ETD Collection for Fordham University. Paper AAI3222437. http://fordham.bepress.com/dissertations/AAI3222437

Tantam, D. (2000). Psychological disorder in adolescents and adults with Asperger syndrome. *Autism, 4*(1), 47-62.

Tantam, D. (2003). The challenge of adolescents and adults with Asperger Syndrome. *Child and Adolescent Psychiatric Clinics of North America, 12,* 143-163.

Tonge, B. J., Brereton, A. V., Gray, K. M., & Einfeld, S. L. (1999). Behavioral and emotional disturbance in high-functioning Autism and asperger syndrome. *Autism, 3*(2), 117-130.

U.S. Department of Education. (2008). *Twenty-five years after A nation at risk.* Retrieved from: http://www2.ed.gov/rschstat/research/pubs/accountable/accountable.pdf

Ups and downs of American STEM Education. (2012). Retrieved from http://blog.stemconnector.org/ups-and-downs-american-stem-education

VanBergeijk, E. O., Klin, A., & Volkmar, F. R. (2008). Supporting more able students on the Autism spectrum: College and beyond. Journal of Autism and Developmental Disabilities, *38*(1), 1359-1370.

Volkmar, F. R., & Klin, A. (2005). Issues in the classification of Autism and related conditions. In F. Volkmar, R. Paul, Klin Ami & D. Cohen (Eds.), *Handbook of Autism and pervasive developmental disorders* (pp. 5-41). Hoboken, NJ: John Wiley and Sons, Inc.

Wei, X., Christiano, E. R., Yu, J. W., Blackorby, J., Shattuck, P., & Newman, L. A. (2013). Postsecondary pathways and persistence for STEM versus non-STEM majors: Among college students with an Autism Spectrum Disorder. *Journal of Autism and Developmental Disorders*, DOI: 10.1007/s10803-013-1978-5

Wheland, E. R., Donovan, W. J., Dukes, J. T., Qammar, H. K., Smith, G. A., & Williams, B. L. (2013). Green action through education: A model for fostering positive attitudes about STEM. *Journal of College Science Teaching, 42*(3), 46-51.

Wing, L. (1981). Asperger's syndrome. *Psychological Medicine, 11,* 115-129.

Wing, L. (1993). The definition and prevalence of Autism. *European Child and Adolescent Psychiatry, 2*(2), 61-74.

Wynn, T., & Harris, J. (2012). Toward a STEM + arts curriculum: Creating the teacher team. *Art Education, 65*(5), 42-47.

In: S.T.E.M. Education
Editor: Satasha L. Green

ISBN: 978-1-63117-806-1

Chapter 7

TEACHING CHILDREN AND ADOLESCENTS WITH TRAUMATIC BRAIN INJURY IN INCLUSIVE STEM CLASSROOM SETTINGS

***Cheryl A. Utley*[1] *and Festus E. Obiakor*[2]**
[1]Chicago State University, US
[2]Valdosta State University, US

ABSTRACT

Persons with disabilities are underrepresented in the science and engineering workforce compared to the population as a whole (National Science Foundation, 2013). Although traumatic brain injury (TBI) is a significant health problem in the United States, the focus is on providing children and adolescents with TBI with opportunities to be placed in the general education classrooms at the earliest possible point. This chapter (a) describes causes, symptoms, and challenges following TBI (e.g., physical, emotional, and cognitive difficulties), (b) distinguish mild TBI from other mild categories of disability, and (c) identify classroom interventions and strategies. Accommodations to educational programs in science, technology, engineering and math (STEM) programs are discussed.

INTRODUCTION

Collectively, students with disabilities are underrepresented in science, technology, engineering, and math (STEM) programs and, later, as adults with disabilities in STEM professions (Melber & Brown, 2008; Moon, Todd, Morton, & Ivey, 2012). In the biennial report *Women, Minorities, and Persons with Disabilities in Science and Engineering*, as mandated by the Science and Engineering Equal Opportunities Act (Public Law 96-516), persons with disabilities are underrepresented in the science and engineering workforce compared to the population as a whole (National Science Foundation, 2013). This underrepresentation has been attributed to a variety of factors such as (a) informal exposure to science educational experiences, (b) lack of access to informal education, (c) a lack of elementary level teacher training in accommodations and inclusive classroom techniques in STEM, and (d) few educational opportunities in math, science and engineering fields as career options. Mostly recently, the American Association for the Advancement of Science (AAS) (2002) noted that although the U.S. has made important progress in helping students with disabilities reach their fullest potential, much more remains to be done for these students to pursue technical careers in STEM, pre-college and post-secondary education, and competitive employment. To change this situation, a number of legislative mandates (e.g., IDEA, ADA) have been enacted to provide access to a quality education for individuals with disabilities, specifically students with traumatic brain injury (TBI) through the provision of classroom accommodations.

This chapter (a) provides an overview of the causes, symptoms, and challenges (e.g., physical, emotional, and cognitive difficulties) facing students with TBI, (b) distinguishes mild TBI from other mild categories of disability, (c) identifies classroom interventions and strategies, and (d) identifies resources and services for educators, professionals, and parents.

Defining Traumatic Brain Injury

Traumatic Brain Injury (TBI) is a serious health problem in the United States. Traumatic brain injuries that are not the result of a medical condition are usually the result of accidents involving 4-wheelers or motor vehicles/traffic accidents, sport injuries, hunting accidents, and/or injuries from falling. National statistics indicated that approximately 1.7 million people receive traumatic brain injuries every year. Of children 0-19 years old, TBI results in 631,146 trips to the emergency room annually, 35,994 hospitalizations, and nearly 6,169 deaths (National Center for Injury Prevention and Control, 2012; Office of Special Education Programs, 2012). During the years 2002-2006, trends in the estimated average annual rates of TBI-related emergency department visits, hospitalizations and death by age group showed that (a) older adolescents, and adults ages 65 years and older are more likely to sustain a TBI; (b) rates are highest among children, ages 0 to 4, and ages 75 and older; (c) in every age group, TBI rates are higher for males than females; and (d) falls are the leading cause of TBI (Faul, Xu, Wald, & Coronado, 2010).

For children and adolescents, ages 6-21, *The Individuals with Disabilities Education Act* [IDEA] (1997) required public schools to provide children with disabilities a free, appropriate public education in the least restrictive environment appropriate to their individual needs. IDEA defines traumatic brain injury (TBI) as:

> An acquired injury to the brain caused by an external physical force, resulting in total or partial functional disability or psychosocial impairment, or both that adversely affects a child's educational performance. The term applies to open or closed head injuries resulting in impairments in one or more areas, open or closed head injuries resulting in impairments in one or more areas, such as cognition; language; memory; attention; reasoning; abstract thinking; judgment; problem-solving; sensory, perceptual, and motor abilities; psychosocial behavior; physical functions; information processing; and speech. The term does not apply to brain injuries that are congenital or degenerative, or brain injuries induced by birth trauma. (34 C.F.R., Sec. 300.7[6] [12])

Currently, the focus is on providing children and adolescents with TBI with the opportunities to be placed in the general education classrooms at the earliest possible point (Blosser & DePompei, 1991; Klomes, 2000; Martin, 1991). Nationally, children and adolescents with TBI have access to comprehensive educational services. Data from the 31st Annual Report to Congress on the Implementation of IDEA (2009) have revealed several important facts regarding the percentage of students with TBI, ages 6-21, served by the disability category of TBI in general education classrooms. In the year 2007, 43.9% of students served under the category of TBI were educated inside the general education class; 24.7% of the students were served in general education classes no more than 79% of the day and no less than 40% of the day; and 22.5% of the students were served less than 40% of the day in general education classes; and 8.8% of the students were served in other school environments (e.g., special school, homebound).

Causes and Symptoms of TBI

Savage (2001) noted that there are many different terms used interchangeably to describe brain injury such as TBI, acquired brain injury (ABI), head injury, concussion, and shaken baby syndrome. All brain injuries are acquired, through traumatic or non-traumatic mechanisms.

Traumatic brain injuries result from external causes and are divided into 2 categories: (a) closed brain injuries are a result of a non-penetrating blow to the head; and (b) open brain injuries are when the skull has been crushed or seriously fractured. Closed brain injuries may result in severe injuries, cause shearing, bruising, bleeding of the brain, or even stretching/tearing of axons as the brain scrapes against the bony ridges of the "brain vault", and include non-penetrating blows to the head or violent shaking as occurs in whiplash or shaken baby syndrome. Open brain injuries may result in a crushed or seriously fractured skull, prolonged hospitalization and extensive rehabilitation, and incomplete rehabilitation and poor prognosis for returning to pre-injury status. These injuries typically require additional treatment such as removal of bone fragments or closure of the wound and often require a long period of rehabilitation.

By contrast, non-traumatic brain injuries (also referred to as acquired brain injuries) may result from internal causes such as cerebral vascular accidents (i.e., stroke), vascular occlusions (i.e., atherosclerosis), hemorrhagic or embolic strokes, rupture of arteriovenous malformations (AVM) brain tumors, hypoxia/anoxia (i.e., lack of adequate oxygen), infections of the brain, brain abscesses, meningitis (outer brain layers), encephalitis (inner brain layers), metabolic disorders (e.g., kidney, liver disease), ingestion of toxic substances, inhalation of organic solvents, ingestion of heavy metals (e.g., mercury, lead), alcohol and drug abuse. Ghajar (2000) and Senelick and Dougherty (2001) noted that TBI is graded as mild, moderate, or severe on the basis of the level of consciousness or Glasgow coma scale (GCS) score after resuscitation (panel).

Mild TBI (GCS 13–15) is in most cases a concussion and there is full neurological recovery, although many of these patients have short-term memory and concentration difficulties. In moderate TBI, (GCS 9–13) the patient is lethargic, and in severe injury (GCS 3–8) the patient is comatose, unable to open his or her eyes or follow commands. The consequences of a brain injury can be diffuse and will vary depending on the location and the severity of the injury. Complications can include motor, sensory, and cognitive impairments, feeding disorders, and communication issues. After the child is stabilized, the rehabilitation process begins – frequently while the child or adolescent is still in a coma.

DISTINGUISHING TBI FROM OTHER MILD DISABILITY CATEGORIES

Intellectual disabilities (e.g., mild mental retardation) are less frequently identified in children with mild TBI (MTBI). Children and adolescents with MTBI are more similar to students with learning disabilities (LD), who have deficits in lower level memory and/or perceptual skills that are associated with specific academic impairments in verbal and numerical literacies, fine-motor skills, and social functioning. Both groups of children and adolescents may demonstrate (a) difficulties with impulse control, (b) difficulties with generalizing and integrating skills, (c) inappropriate behavior due to poor social judgment, (d) difficulty with sustained attention, and (e) a need for memory aids due to difficulties with short term memory (Michaud, Semul-Concepcion, Duhaime, & Lazar, 2005; Shapiro, Church, & Lewis, 2002).

Distinguishing characteristics of children and adolescents with MTBI from LD include (a) specific impairments in the areas of attention/concentration, information processing speed, memory, impulsiveness and mood swings; (b) sudden onset due to external events that are precipitated by a period of normal, uncompromised development; (c) loss of consciousness, hence, clear evidence of neurological damage; (d) marked pre-post-injury contrast in cognitive, behavioral, physical capabilities; (e) requires emergency medical care or extended hospitalization; (f) paresis, paralysis, and spasticity require specialized treatment; (f) distractibility is provoked by internal and external stimuli; (g) complex array of neurobehavioral complications require modified and intensive applications of instructional and management techniques; and (h) moderate to severe problems with memory, new learning, and speech/language problems.

Similarities and differences also exist in children and adolescents with MTBI and emotional disturbance (ED). Both groups of children and adolescents may demonstrate the following characteristics: (a) inappropriate behaviors in social settings, (b) an inability to maintain relationships with peers and teachers, (c) a general pervasive mood of unhappiness or depression, and (d) difficulty with circumstances that are unanticipated. Hibbard, Gordon, Martin, Raskin, and Brown (2001) noted that children with MTBI have the following distinguishing characteristics: (a) marked contrast between pre-and post onset capabilities; (b) reduced processing speed, memory loss, impaired executive functions; (c) pre-injury learning is largely intact; (d) loss of balance, paralysis, visual/ sensory changes, headaches; (e) peer interactions affected by cognitive and behavioral difficulties, and reduced social skills.

Assessment of TBI in Children and Adolescents

There are numerous evaluations and tests that can be conducted to help determine at what educational level the student is functioning and which areas or skills have been most affected

(Deaton, 1999; Ylvisaker, Chorazy, Feeney, & Russell, 1999). Neuropsychological tests are often used with children and adolescents with brain injury because they focus on how the brain functioning affects the student's behavior and skills. Neuropsychological testing is characterized by the depth of attention paid to basic brain functions that may impact more complex skills.

Typically, a neuropsychological evaluation includes measures of attention, memory, expressive and receptive language, fine motor skills, perceptual-motor skills, problem solving, and abstract reasoning. A neuropsychological evaluation also includes assessment of a child's non-test specific behavior, approach to testing, and social skills. Each of these areas is evaluated in multiple ways and with attention paid to both the child's strengths and his/her weaknesses.

To determine the present level of the student's functioning, a comprehensive battery of tests must be administered across the following areas: (a) academic (e.g., reading, math, science, writing, career education), (b) intellectual (e.g., intelligence, learning process, learning style, and learning ability), (c) social-emotional development (e.g., distractibility, impulsivity, attention span, aggression, withdrawal, acting out, immaturity/inadequacy), (d) adaptive behavior (e.g., history of developmental milestones, dressing, eating, personal hygiene, independent living), (e) motor development (e.g., fine motor, gross motor, sensory-motor integration, mobility, and muscular control), (f) language development (e.g., language development, pragmatic language, speech production, articulation), (g) sensory (e.g., vision and hearing), and (h) vocational skills (e.g., entry level work skills, occupational interests, work attitudes, job seeking skills, and job keeping skills).

EDUCATIONAL CHALLENGES FACING STUDENTS WITH TBI

First, children and adolescents experience varying physical challenges post TBI – most commonly fatigue, headaches and dizziness. Fatigue, in particular, is a major contributor to behavioral challenges, as it undermines their ability to deal with emotional and cognitive issues that are commonly found post TBI. Second, the majority of children and adolescents experience significant emotional challenges after TBI, including depression, anxiety, difficulty in accepting changes in oneself and problems in controlling the expression of emotions. Emotional reactions to brain injury (for example, sadness, self-blame, feelings of worthlessness, nervousness, etc.) can directly lead to behaviors that are counterproductive within the classroom. Cognitive deficits, the third area of challenge, also may lead to behavioral issues. For example, a common problem for children and adolescents after TBI is memory difficulties. They may lack motivation because they neglect their homework or fail to follow-up on assignments. Impaired memory skills, instead, may underlie these behaviors.

The fourth issue, developmental challenges, refers to brain injury's having a differing impact on a student's functioning, depending upon the child's age at injury. For example, students who are injured in adolescence will have already developed a sense of identity prior to injury. In the context of TBI, changes in the self will be more emotionally upsetting for these adolescents than for younger children, with a less well developed self-image. Developmental challenges also refer to the fact that the impact of injury shifts as the student is confronted with increasingly complex age-appropriate tasks. For example, children and adolescents with TBI may easily handle simple arithmetic problems only to experience increasing frustration and failure as math becomes more cognitively complex. Developmental challenges also appear in social contexts – as the 'rules' of social living change in the adolescent years. As children and adolescents with TBI fail to learn increasingly complex social 'rules', they may feel more alienated in social situations.

Finally, children and adolescents with TBI face developmental challenges based on the degree to which (before or after injury) the developmental tasks of each phase of childhood have been achieved. For example, a student injured at age five experiences greater challenges, in that foundations for *all* subsequent learning – learning to listen, question and engage socially – need to be established after brain injury. Failure (because of TBI) in this early-childhood developmental task will have a proportionally stronger effect on learning that follows.

While the majority of children and adolescents with TBI return to school, their educational and emotional needs are likely to be very different from what they were prior to the injury. Although children and adolescents with TBI may appear to function similar to children identified with other disabilities, it is important to recognize that the sudden onset of a severe disability resulting from trauma is very different from the onset of a disability from birth (Denton, 2008). Children and adolescents with TBI can often remember how they were before the trauma, which can result in a constellation of emotional and psychosocial problems not usually present in students with congenital disabilities (Hunt, Lees, Philbrick, Richard, Guerrier, & Hammond, 2006). Furthermore, trauma impacts family, friends, and professionals who recall what children and adolescents were like prior to injury and who have difficulties in shifting and adjusting their own personal goals and expectations.

Hibbard et al. (2001) presented four facts that are important to remember about long-term changes in students' behaviors following TBI: (1) rates of recovery and behaviors vary greatly from student to student; (2) changes are unlikely to disappear fully over time; the student's recovery will most likely only be partial; (3) negative consequences may not be seen immediately but only emerge when developmental demands reveal deficits and problems; and (4) an injured brain is less likely to meet the increasingly complex tasks all children and adolescents face as they get older.

Misclassification Issues

There are many ways in which children and adolescents with TBI can be misclassified in the educational process. Factors influencing the classification and placement of children and adolescents with TBI in general and special education classrooms include but are not limited to: (a) children and adolescents returning back to school following a severe or moderate TBI with their learning difficulties unrecognized, typically due to poor transition services between hospitals and schools. In such cases, hospitals fail to provide schools adequate information about the child's cognitive, behavioral or other challenges and guidance as to accommodations that should be implemented to effect a smooth transition between hospitals and schools; (b) children and adolescents with mild TBI can be sent home from the emergency room or physician's office without parents knowing that their child has an injury; (c) the child's school does not learn that an injury has occurred and children and adolescents resume their regular classes without curriculum accommodations; (c) children and adolescents with TBI who experience residual problems and the effects of TBI are left unrecognized, then student performance is attributed to other causes, such as poor motivation; (d) traditional approaches to psycho-educational assessment fail to provide necessary insight as to how cognitive deficits (e.g., attention, processing speed, memory, etc.) following TBI affect academic performance; and (e) when undergoing a traditional psycho-educational evaluation, children and adolescents with TBI are often found to perform in the average range on tests of intelligence and academic achievement, and therefore, are judged to be normal and not requiring special education services.

School Reentry and Transition Issues

Blosser and DePompei (1991) observed that children and adolescents with TBI and their families often encounter difficulties and barriers at the time of school reentry. These difficulties include communication problems between rehabilitation professionals and educational professionals. Differences in the orientation of medical, rehabilitation, and educational professionals may result in obstacles preventing children and adolescents with TBI with (a) adequate program coordination, (b) access to counseling, (c) use of appropriate evaluation procedures, (d) developing diagnostic and testing schedules, and (e) having expectation standards.

Approximately two decades ago, several scholars identified problems impacting hospital-school transition issues for children and adolescents with TBI. Public schools may lack appropriate services and professionals who are trained in planning and providing educational programs and medical services for children and adolescents with TBI (Cohen, 1986, 1991; Cooper, 1986; Cooley & Singer, 1991; Council for Exceptional Children, 1987; Lehr, 1990; Lindsey, 1981; Martin, 1988, 1991). Flexible programs and placement in general and special education classrooms should be handled as a process, monitored on an ongoing basis. Adaptations to programs such as homebound, hospitals, and therapy programs must be considered as options for reentry into schools. Lastly, family expectations for the school return may affect the school reentry and transition process (Klomes, 2000).

With these educational challenges and issues in mind, Klomes (1995, 2000) recommended the following program components pertaining to TBI and school reentry: (1) the designation of a case manager, (2) parent education programs that offer information and calendar activities, (3) individualized education programs (IEP) that plan out a child's school program and related services, (4) staff training sessions for school professionals, (5) multidisciplinary collaboration and conferences, (6) assessment and identification by using appropriate neurological instruments administered by professional trained in neurological testing, (7) open communication between hospital/rehabilitation professionals and educational professionals, (8) integrating students with TBI into general education classrooms, (9) planning flexible programs, (10) careful review and planning of classroom options and placements, (11) adapting instruction materials, (12) eliminating environmental barriers in schools and hospitals, (13) early educational and related services, (14) adjusting differences between hospitals' and schools' educational programming, (15) hospitals and schools defining TBI using similar terminology, (16) careful review of former school records, (17) appropriate special education titles, (18) peer education, and (19) support groups and resources for children, adolescents, and their families.

EFFECTIVE SKILLS OF STEM TEACHERS OF CHILDREN AND ADOLESCENTS WITH TBI

Turkstra and Kennedy (2005) and Ylvisaker, Chorazy, Feeney, and Russell (1999) have identified a set of skills essential to supporting children and adolescents with TBI. These include the following: (a) knowing the consequences of TBI in general and for the specific student with TBI, (b) having a good relationship with the student who has a TBI, (c) challenging the student and giving honest proactive feedback, (d) providing opportunities for success while setting higher and higher expectations, (e) being optimistic about the student's abilities and what the student might be doing in the next few years, (f) finding the positive in the student's behavior and focus on it rather than looking for the negative, (g) flexibility in adapting to the changing needs of the student, (h) being creative and able to think outside the box, (i) finding multiple solutions to problems and encouraging the student to find his own solutions, too (generating multiple solutions to problems can be part of the student's IEP), (j) being enthusiasm and support of the student's progress, and (k) being mature enough to allow the student to grow without creating dependence.

Classroom Interventions in STEM for Students with TBI

Glang, Ylvisaker, Stein, Ehlardt, Todis, and Tyler (2008) noted that "there is very little empirical evidence of effectiveness for interventions to promote positive educational outcomes for children and youth following injury. However, a number of promising practices can be identified from research with children with other disability labels. Although children with TBI often have unique learning and behavioral characteristics, they also have many commonalities with children with other disabilities. This allows for cautious cross-population application of educational and behavioral STEM strategies, which have been validated in relation to specific educational needs rather than general diagnostic categories" (p. 243).

Although learning for many students with TBI is complex in nature, it remains imperative that pedagogy, interventions, and instructional strategies are not inaccessible to these students. Sencibaugh (2008) noted that students with disabilities who have problems accessing, processing and using information presented in STEM classes need specific structures and strategies to assist them in these processes. Content enhancement devices will help students identify, organize, understand, and remember information. In a meta-analysis of research studies on special education content teaching strategies, Scruggs, Mastropieri, Berkeley and Graetz (2010) found a number of specific enhancement devices to be effective: (1) mnemonic strategies that use patterns and associations to aid memory; (2) spatial organizers that use graphic organizers to sort information to aid in understanding and memory; (3) classroom learning strategies that teach students such skills as self-monitoring and study skills; (4) computer-assisted instruction that uses technology to promote drill and practice, simulation and other skills; (5) peer mediation that uses cooperative grouping for mutual academic benefits; (6) study aids that make use of enhanced printed supports such as study guides and text outlines; (7) hands-on or activity-oriented learning that teach concepts via hands-on learning activities; and (8) explicit or direct instruction that includes task analyzed learning and guided and independent student practice.

According to Glang et al. (2008), the empirical support for interventions includes (a) evidence of the effectiveness of component features; (b) studies that have investigated school-based implementations of comprehensive curricula with a range of students; and (c) studies of the use of interventions with students with TBI. In a comprehensive review of literature, these authors found instructional components common to both direct instruction and cognitive strategy instruction effective in teaching children and adolescents with TBI. These components are (a) systematic, explicit instruction and practice, (b) consistent instructional routines, (c) effective task analysis, (d) systematic introduction and modeling of component skills, (e) use of scaffolding/guided practice, (f) rapid instructional pacing, (g) teaching to mastery (criterion referenced instruction), (h) consistently high rates of success, (i) teaching of generalizable strategies, (j) planned and programmed generalization, and (k) frequent and cumulative review.

Instructional Strategies in STEM for Students with TBI

Because each child or adolescent with TBI is unique, there is no one STEM teaching program that applies to all students. By adapting STEM instruction or modifying the STEM environment, however, the children and adolescents with TBI can have greater opportunities for success in a STEM classroom. Earlier, Mira and Tyler (1999) and most recently NICHCY (2012) presented instructional strategies in the areas of receptive and expressive language, attention, impulsiveness, memory, following directions, and motor skills for the inclusion of children and adolescents with TBI in general education classrooms that include STEM subject matter. These are as follows:

Receptive language. Limit the amount of information presented at one time and provide simple instructions for only one activity at a time and use concrete concepts and language. Have students repeat instructions.

Expressive language. Teach students to rehearse silently before verbally replying. Teach students to look for cues from listeners.

Maintaining attention. Provide students with a study carrel or preferential seating. After giving instructions, check for proper attention and understanding by having students repeat them. Teach students to use self-regulating techniques to maintain attention.

Impulsiveness. Teach students to mentally rehearse steps before beginning activity. Reduce potential distractions. Frequently restate and reinforce rules.

Memory. Teach students to use external aids such as notes, memos, daily schedule, and assignment sheets. Provide repetition and frequent review of instructional materials. Provide immediate and frequent feedback to enable students to interpret success or failure.

Following Directions. Provide students with both visual and auditory directions. Model tasks whenever possible. Break multi-step directions into small parts and list them so that students can refer back when needed.

Motor Skills. Have students use a word processor to complete assignments. Allow extra time for completing tasks requiring fine motor skills. Allow students to audiotape the lecture or assign someone to take notes for students during lectures.

Assistive Technology

Assistive technology can also be useful to children and adolescents with TBI when learning STEM content. Some examples of assistive technology include: (a) recording lectures on a tape recorder instead of taking class notes, (b) using books on tape, (c) having students check math work with a calculator, (d) allowing students to use a spell check tool or to type work on the computer, (e) allowing headphones for students who are easily distracted, and (f) using a voice recognition program on a computer to write reports. Prior to selecting assistive technology for students, it is critical that students are evaluated for assistive technology by a specialist.

STEM Curriculum Accommodations

As noted earlier, children and adolescents with TBI experience significant challenges in STEM classrooms. The placement of children and adolescents with TBI in STEM curriculum is often one of the most difficult decisions that have to be made by the general educator. It is recommended that classroom teachers consult with a team of rehabilitation specialists, including a speech therapist and neuropsychologist, as well as specialists in the school system. It may be necessary for persons who are familiar with the child/adolescent's performance at the time of school entry to communicate with representatives from the school and the classroom teacher.

Academic success in a STEM classroom is determined by a number of factors including the interaction between the child's cognitive, emotional, developmental, and physical characteristics, curriculum (e.g., math and science), and school environment (e.g., level of stimulation, classroom pacing, predictability, structure, physical and cognitive demands, and social support) (Hibbard et al., 2001).

Tables 1 and 2 describe mathematics and science curriculum demands, challenges facing students with TBI, and possible solutions.

Table 1. Mathematics Curriculum and Solutions

Curriculum Demands	Challenges of Students with TBI	Possible Solutions
Ability to use syntactic and semantic components of language to solve verbal math problems.	Difficulty with semantic aspects of word problems.	Aid in finding the main idea of the verbal math problem and what information is needed to solve the problem.
Recall and use "math language" when needed. Many complex concepts are carried in a few words (e.g., divide, multiply, add).	Unable to recall the concept associated with a single word. Misses the instruction to "add".	Teach the meaning of single words that carry considerable intent. Aid in recall of the concepts and processes underlying the single word.
Employ sequencing skills to complete a process.	Sequencing skills are often impaired.	Work on meaningful, functional sequencing skills.
Use language to understand the word problem and then complete the math to solve the problem.	Poor recall, inability to find relevance within the word problem. Oftentimes the student can do the math if he/she can understand the words that formulate the problem.	Develop ability to find the main question within the problem and associate the concepts necessary to solve it.

Curriculum Demands	Challenges of Students with TBI	Possible Solutions
Ability to perform basic handwriting skills to integrate the visual images of numbers and shapes.	Poor fine motor ability and inability to integrate visual-motor and visual-cognitive processes.	Work with student to proofread own work; provide concrete cues for abstract math concepts.
Ability to align rows and columns of math problems.	Visual-perceptual problems may inhibit student's ability to align columns and rows.	Utilize visual aids, such as colored lines/rows, templates with windows, graph paper, etc.
Use calculator to complete math calculations.	Students with poor visual memory have difficulty using a calculator.	Utilize repetition and organize information into small units; list steps of process.

Source: Brain Injury Association of Virginia. (2005). *Brain injury and the schools: A guide for educators*. Available online at: www.biav.net.

Table 2. Science Curriculum and Possible Solutions

Curriculum Demands	Challenges of Students with TBI	Possible Solutions
Ability to handle and manipulate small objects.	Decreased ability to pick up or manipulate small objects.	Build up handles, smaller objects with tape/padding; make handles/knobs longer for easier handling.
Awareness of safety concepts.	Inability to recognize safety concepts.	Draw attention to safety issues with bold colors/signs, pair student with a partner to increase awareness.
Ability to maintain attention.	Decreased attention span.	Break projects into smaller steps to hold attention.
Knowledge of concepts such as more than/less than; before/after; when/then.	Inability to recognize relationships and concepts that are not concrete in nature.	Use visual aids and concrete objects to demonstrate relationships.
Recognition of cause and effect.	Inability to understand the relevance of cause and effects.	Devise activities which utilize concrete objects to assist in the understanding of abstract concepts.

Source: Brain Injury Association of Virginia (2005). *Brain injury and the schools: A guide for educators*. Available online at: www.biav.net.

Functional Behavioral Assessment

Children and adolescents with TBI are vulnerable to significant behavioral challenges due to the high incidence of frontal lobe involvement. It is often more effective to redirect inappropriate behavior on the part of children and adolescents with TBI than to discuss it with them, particularly within the context of a classroom situation. Children and adolescents with TBI may not fully understand comments made by educators concerning their behavior. It is also important not to draw attention to misbehavior since this may prove reinforcing or may prove embarrassing to children and adolescents with TBI.

The key to effective behavioral analysis is to look at what motivates the behavior and consider what environmental factors are promoting the behavior. It is important when developing a behavior plan to look at ways to prevent inappropriate behaviors and what skills students need to learn to avoid the behavior. It is important when developing a behavior plan to look at ways to prevent behaviors from happening and what skills the student needs to learn to avoid the behavior. Children and adolescents with TBI often lack the ability to foresee the consequences of their actions and they also have difficulties controlling their impulsive behavior. Children and adolescents with TBI whose behavior interferes with their learning or the learning of the other students should have a behavior plan. The individualized educational plan (IEP) team should meet and complete a functional behavioral assessment (FBA). Best practice for a FBA should involve all the IEP team members in a brainstorming session to discuss the specific behaviors of children and adolescents with TBI so that all the IEP team members agree on inappropriate student behaviors. The IEP team should only look at one or two behaviors at a time. The next step is to gather data including: duration, frequency, and intensity of the behavior. The IEP team should collect data on the actual behaviors, interview parents and other individuals about the history of the behavior and if it happens in other settings, and review school records. Important questions that must be addressed are (a) when, where and with who does the behavior most often occur? (b) when, where and with whom does the behavior seldom, if ever occur? (c) what happened just before and after the behavior? (d) what did the school staff do in response to the behavior? (e) is this a "hidden" reward? (f) what are the environmental factors involved in the students behavior? (g) are there any health or medical factors related to the behavior? (i.e., change in medication, illness, dental issues, etc.)? (h) are there any factors at home related to the behavior? (i.e., death, divorce, etc.), (i) why does the team think the student is doing the behavior? and (j) what need is the behavior fulfilling for the student?

A positive behavior support plan (PBSP) is a pro-active way to address a specific behavior that incorporates positive strategies and supports designed to increase appropriate replacement behaviors. It should include two parts. The first part is a list of things the school can do to prevent or decrease the likelihood of the behavior occurring. The second part is to look at what skills the student with TBI needs to learn in order to get their needs met in an appropriate way. It is very difficult to change a behavior if you have not helped them get their need met in an appropriate way. A good PBSP also includes a (a) plan to phase out intervention techniques, (b) schedule of follow up meetings to determine if the plan is successful and to make changes, (c) discussion of how the plan will be implemented outside of the school setting, and (d) how the plan relates to goals on the student's IEP.

Key Elements of STEM Programs for Children and Adolescents with TBI

The AAAS launched a unique internship program, called ENTRY POINT! demonstrating that students with disabilities have enormous career potential. ENTRY POINT! is a collaborative project involving AAAS, government agencies, and industry. The program identifies, screens, and places undergraduate and graduate students with disabilities who are pursuing careers in STEM in paid summer internships. Key elements of the project include the following: (1) full access to the Internet, computers, software programs, assistive technology and other resources; (2) family and community support for students with disabilities who pursue science and engineering degrees and transition from campus to the competitive workplace; (3) mentors and role models who can break down barriers and encourage students with disabilities to persevere in STEM; and (4) opportunities that provide professional internships so they can gain confidence in their own intellectual abilities and skills.

Recommendations to educators and policy makers include: (1) protecting and strengthening the laws that have been enacted; (2) encouraging businesses, educators, and health care providers to support technology (e.g., assistive technology) that can foster independence; (3) providing legislative incentives to encourage corporate internships and the hiring of persons with disabilities; (4) improving research on students with disabilities and their progress in a variety of fields, particularly STEM; and (5) encouraging communities, businesses, and schools to include persons with disabilities in local organizations.

Conclusion

Over these past few years, the field of TBI has experienced tremendous growth. Much work still needs to be done to understand the complexities of TBI. This chapter has discussed a number of pertinent issues to the teaching of children and adolescents with TBI in inclusive STEM classrooms. These issues include but are not limited to: causes and symptoms of TBI, educational challenges for children and adolescents with TBI, and effective STEM teaching strategies for these individuals to be successful in general and special education classrooms.

References

American Association for the Advancement of Science. (2002). *New Career Paths for Students with Disabilities*. Washington, DC.

Altimer, L. (1992). Pediatric central neurologic trauma: Issues for special patients. *AACN Clinical Issues in Critical Care Nursing, 3*(1), 31-4.

Bigler, E. (1988). Acquired cerebral trauma: Epilogue. *Journal of Learning Disabilities, 21*(8), 476-485.

Blosser, J., & Depompei, R. (1989). The head injured student returns to school: Recognizing and treating deficits. *Topics in Language Disorders, 9*(2), 67-77.

Brain Injury Association of Virginia. (2005). *Brain injury and the schools: A guide for educators*. Retrieved from: www.baiv.net

Carney, J., & Gerring, J. (1990). Return to school following severe closed head injury. A critical phase in pediatric rehabilitation. *Pediatrician, 17*, 222-229.

Carter, R., & Savage, R. (1985). Education and the traumatically brain injured: Rights, protections, and responsibilities. *Cognitive Rehabilitation, 3*(5), 14-17.

Center for Disease Control. (2010). *Traumatic brain injury in the United States: Emergency department visits, hospitalizations and deaths, 2002–2006*. Retrieved from: http://www.cdc.gov/traumaticbraininjury/pdf/blue_book.pdf

Cohen, S. (1986). Educational reintegration and programming for children with head injuries. *Journal of Head Trauma and Rehabilitation, 1*(4), 22-29.

Cohen, S. (1991). Adapting educational programs for students with head injuries. *Journal of Head Trauma 6*(1), *56-63*.

Cook, J. (1991). Higher education: An attainable goal for students who have sustained head injuries. *Journal of Head Trauma Rehabilitation, 6*(1), 64-72.

Cooley, E., & Singer, G. (1991). On serving students with head injuries: Are we reinventing a wheel that doesn't roll? *Journal of Head Trauma Rehabilitation, 6*(1), 47-55.

Cooper, D. (1986). Special needs: Education in a hospital setting. *ERS Spectrum, Winter*, 39-43.

Council of Exceptional Children. (1987). The regular education initiative. *Journal of Learning Disabilities, 20*(5), 289-293.

Deaton, A. (1999). *Cumberland hospital's pediatric brain trauma guide for families* (3rd ed.). New Kent, VA: Cumberland Hospital for Children and Adolescents.

Denton, G. L. (2008). *Brain Lash: Maximize Your Recovery from Mild Brain Injury* (3rd ed.). New York, NY: Demos Medical Publishing.

Dixon, T., Goll, S., & Stanton, K. (1988). Case management issues and practices in head injury rehabilitation. *Rehabilitation Counseling Bulletin, 31*(4), 325-343.

Faul, M., Xu L., Wald, M. M., & Coronado, V. (2010). Traumatic brain injury in the United States: Emergency department visits, hospitalizations and deaths, 2002-2006. Atlanta, Georgia: Centers for Disease Control and Prevention, National Center for Injury Prevention and Control.

Faurve, M. (1988). Including young children with "new" chronic illness in an early childhood education setting. *Young Children, 43*, 71-77.

Ghajar, J. (2000). Traumatic brain injury. *Lancet, 356*, 923-929.

Glang, A., Ylvrisaker, M., Stein, M., Ehlhardt, L., Todis, B., & Tyler, J. (2008). Validated instructional practices: Application to students with traumatic brain injury. *Journal of Head Trauma Rehabilitation, 23*(4), 243-251.

Hall, D., & DePompei, R. (1986). Implications for the head injured reentering higher education. *Cognitive Rehabilitation*, 6-8.

Hibbard, M., Gordon, W.A., Martin, T., Raskin, B., & Brown, M. (2001). *Students with traumatic brain injury: Identification, assessment, and classroom accommodations.* Research and Training Center on Community Integration of Individuals with Traumatic Brain Injury, Mount Sinai School of Medicine, NY.

Hunt, L. G., Lees, C., Philbrick, P., Rickard, S., Guerrier, T., & Hammond, F. (2006). *Brain Injury: It is a journey a practical guide for families.* Wake Forest, NC: Lash & Associates Publishing/Training Inc.

Individuals with Disabilities Education Act (IDEA). (1997).U.S.C., 2034 C.F.R., Sec. 300.7[6] [12].

Klomes, J. M. (1995). *A study of the components associated with the school reentry process for students with traumatic brain injury*. Unpublished doctoral dissertation. Northern Illinois University.

Klomes, J. M. (2000). The school reentry process for students with traumatic brain injury. In F. E. Obiakor, S. A. Burkhardt, A. F. Rotatori, & T. Wahlberg (Eds.), *Intervention Techniques for Individuals with Exceptionalities in Inclusive Settings* (Vol. 13) (pp. 199-216), *Advances in special education*, Stamford, CT: JAI.

Lehr, E. (1990). *Psychological management of traumatic brain injuries in children and adolescents*. Rockville, MD: Aspen Publication.

Lindsey, C. (1981). The classroom teacher and the hospitalized child. *Education Unlimited,* 30-32.

Martin, R. (1988). Legal challenges in educating the traumatic brain injured students. *Journal of Learning Disabilities, 21*(8), 471-475.

Martin, R. (1991). Special education legal update. *Carle Center for Health, Law, and Ethics. Winter.*

McClelland, M., & Sands, R. (1991). *The missing voice in interdisciplinary communication.* Philadelphia, PA: University of Pennsylvania, Division of Psychology.

Melber, L., & Brown, K. (2008). Not like a regular science class: Informal science education for students with disabilities. *The Clearing House, 82*(1), 35.

Michaud, L., Semel-Concepcion, J. Duhaime, A., & Lazar, M. (2002). Traumatic brain injury. In M. L. Batshaw (Ed.), *Children with Disabilities* (5th ed.) (pp. 525-545). Baltimore, MD: Paul H. Brooks Publishing Company.

Mira, M. P., & Tyler, J. S. (1991). Students with traumatic brain injury: Making the transition from hospital to school. *Focus on Exceptional Children, 23*(5), 1-12.

Mira, M. P., & Tyler, J. S. (1999). *Traumatic brain injury in children and adolescents: Source book for teachers and other school personnel.* Austin, TX: Pro-Ed Publishing.

Moon, N.W., Todd, R.L., Morton, D.L., & Ivey, E. (2012). Accommodating students with disabilities in science, technology, engineering, and mathematics (STEM): Findings from research and practice for middle grades through University Education. Atlanta, GA: Center for Assistive Technology and Environmental Access College of Architecture.

National Association of State Boards of Education (NASBE). (1992). *Winners all: A call for inclusive schools.* Virginia: NASBE.

National Center for Injury Prevention and Control. (2012). *Traumatic brain injury.* Retrieved from: http://www.cdc.gov/TraumaticBrainInjury/index.html

National Head Injury Foundation Flyer. (1992). *Mission Statement Flyer.* Brookfield, IL: Head Injury Association Inc. Chapter of NMIF.

National Science Foundation (2013). *Women, minorities, and persons with disabilities in science and engineering*: Available on line at: http://www.nsf.gov/statistics/wmpd/2013/pdf/nsf13304_digest.pdf

Office of Special Education Programs Data Accountability Center. (2012). *Data tables for OSEP state reported data.* Retrieved from: https://www.ideadata.org/TABLES35TH/B1-2.xls.

Savage, R. (July, 2001). *An analysis of 15, 024 children with traumatic brain injury.* Presentation at the 20th Annual Symposium on the Building Blocks of Change, Brain Injury Association of America, Atlanta, GA.

Scruggs, T.E., Mastropieri, M.A., & Berkeley, S., & Graetz, J. (2010). Do special education interventions improve learning of secondary content? A meta-analysis. *Remedial and Special Education,* 36, 437-449.

Sencibaugh, J.M. (2008). A synthesis of content enhancement strategies for teaching students with reading difficulties at the middle and secondary level. Retrieved from. http://www.redorbit.com/news/education/1538415/asynthesis_of_content_enhancement strategies_for_teaching_studentswith/disabilities html.

Senelick, R., & Dougherty, K. (2001). *Living with brain injury: A guide for families.* Birmingham, AL: HealthSouth Press.

Shapiro, B., Church, R., & Lewis, M. (2002).*Children with Disabilities.* Baltimore: Paul H. Brooks Publishing.

Turkstra, L., & Kennedy, M. (2005). Evidence-based practice for cognitive-communication disorders after traumatic brain injury. *Seminars in Speech Language, 26*(4), 213-214.

Ylvisaker, M., Chorazy, A., Feeney, T., & Russell, M. (1999). Traumatic brain injuries in children and adolescents: Assessment and rehabilitation. In Rosenthal, M., Griffith, E.,

Kreutzer, J., & Pentland, B. (Eds.).*Rehabilitation of the adult and child with traumatic brain injury* (3rd ed., pp. 356-392). Philadelphia: F. A. Davis.

Ylvisaker, M., & Feeney, T. (1998). Everyday people as supports: Developing competencies through collaboration. In. M. Ylvisaker (Ed.), *Traumatic brain injury: Children and adolescents* (pp. 429-445). Newton, MA: Butterwort-Heinemann.

In: S.T.E.M. Education
Editor: Satasha L. Green

ISBN: 978-1-63117-806-1

Chapter 8

WORKING WITH LEARNERS WITH PHYSICAL AND HEALTH IMPAIRMENTS IN STEM

*Amina M. Turton**
Vancouver Island University, Canada

ABSTRACT

This chapter addresses the issues surrounding the increased access of science, technology, engineering and mathematics (STEM) for students with physical and health impairments. Descriptions of the breadth of physical and health impairments that are currently being served under the Individuals with Disabilities Act 2004 (IDEA, 2004) within our schools, homes and hospitals are discussed. Focus is given to STEM approaches that have been developed as a response to the decline in K-12 students and graduates within the United States STEM 'pipeline'. This chapter further looks at how these approaches are suitable to the teaching of those individuals with physical and health impairments.

INTRODUCTION

Chronic physical illness in children and adolescents may greatly disrupt child and family life as well as the schooling experience. Missing school can lead to problems in keeping up with schoolwork and in social relationships, which can have long-term consequences such as poor self esteem and socialization skills (Sturge, Garralda, Boissin, Dore, & Woo, 1997). In 2004, 8.4% of student had other health impairments (US Department of Education, 2005). Physical disabilities are considered and subsumed under the 2004 Individuals with Disabilities Education Act (IDEA) of the special education category Orthopedic Impairment (OI). Under IDEA health impairments are considered Other Health Impaired (OHI). The lines between both groups are not well defined and are at some junctures quite blurred; for example the chronicity of a physical impairment and/or HI may well develop into an OHI and vice

* Amina.Turton@viu.ca

versa. Long-term support either through medicinal treatment, physical therapy or both may have a considerable impact on the attendance, participation/engagement and successful completion of school. The domino effect this has on school completion, post secondary educational opportunity and employment can be observed through both the review of historical data and the current understanding of how those individuals fare within our educational systems. According to Rovert, Erhlich, and Hoppe (1988), one's education and the opportunities it affords can be severely compromised with chronic health conditions. Though these authors focused on Type I Diabetes Melitus they delved deeper than the typical sugar levels and social/emotional aspects that had been traditionally researched; their research revealed that the neuro-cognitive development, the visio-spatial and verbal areas of cognition are also affected. This awareness means we must as fields come together to best support an individual living with these issues.

We now understand that what encompasses the sphere of child's health and development are the result of a unique "dynamic interplay" of molecular biology (physiology), personal experience, and community context. This perspective can result in a more holistic view and experience of what it means to live with an OHI. Understanding what leads individuals to be at a higher risk for both PI and HI means incorporating the 'dynamic' within the clinical equation that has historically omitted holistic factors. We know that a large part of our wellbeing is correlated with our environments. This piece can set the stage for positive supported quality of life outcomes or its opposite (McPherson, Arango, Fox, Lauver, McManus, Newacheck, Perrin, Shonkoff, & Strickland, 1998; Woodrich & Spencer, 2008). A significant part in the well-being and quality of life are the environments that place the child either at risk which can negatively affect both physical and mental development or set the stage for the predisposition/risk ration (McPherson, Arango, Fox, Lauver, McManus, Newacheck, Perrin, Shonkoff, & Strickland, 1998). Woodrich and Spencer, (2008) specified in their study that the role of education may change in relation to the increased prevalence and diagnoses of health impairments within our school-aged population. They go so far as to say that the bio-medical stance that has traditionally been used as the structure for the identification of OHI and PI now works in favor of their practice allowing the psychologist to share knowledge that leads to appropriate classroom support and class-hospital - collaboration.

Special Education, PI and HI

Broadening the knowledge scope for special education teachers so that they can enhance their capacities in this realm and make appropriate plans while gaining deeper understanding of various OHI or PI seems like an appropriate step to take. The changes necessary would need to occur at the individualized, local, and institutional levels (Nasir & Hand, 2006). This deeper comprehension should be sought after, with the caveat that the complications which can arise from chronic health impairments may include co-occurring cognitive impairments; however this is not true for all situations and the converse is just as likely to occur. Having a perspective that encompasses the former and not the latter can be very detrimental to students, teachers and families (Carter, Cushing, Clark & Kennedy, 2005; McDonnell, Hardman, & McDonnell, 2003).

Underneath the umbrella of OHI we see patterns emerging that speak to us about the prevalence data of different disorders and conditions. We see a rise in Attention Deficit Hyperactivity Disorder (ADHD) and Attention Deficit Disorder (ADD) from the U.S. Department of Education, (2001; 2008) reports. These identifying conditions carry the bulk of OHI diagnoses. Increases in Type I Diabetes Melitus (Holmes, Cant, Fox, Lampert, & Greer, 1999); epilepsy (Goldstein et al., 2004; Huberty, Austin, Huster, & Dunn, 2000), and juvenile cancers (Brown et al., 1998; Buono et al., 1998) are noted which bring with them the challenge to identify with clarity where the role of an educational psychologist/school and pediatrician begin and end. Examples school psychologists encountered outside of the ADHD bubble, that would be considered OHI are conditions such as asthma/pulmonary disease, diabetes, and epilepsy. The data showed a 1:5 ration of psychologists who had experience in the diagnoses and support of individuals outside of AD/H/D diagnoses. What is missing as noted by Wodrich & Schmitt (2003) is that despite getting some spotlight attention in appropriate disease focus; how these conditions manifest and affect classroom performance are remarkably absent from the literature base.

Some school districts in British Columbia, Canada utilize an approach where the family, psychologist, special educator and pediatrician meet monthly and student cases are discussed. Appointments are made as one would in a typical doctor's office though the focus is on a school district and its families (SD 68, BC, Canada, 2013). This model though not used widely and one that may be criticized for its inability to really account for the sheer volume of families and children in need is still at the beginning stages.

Physical Impairments

"Significant limitation, impairment, or delay in physical capacity to move, coordinate actions, or perform physical activities, exhibited by difficulties in one or more of the following areas: physical and motor tasks, independent movement, performing basic life functions. The term shall include severe orthopedic impairments or impairments caused by congenital anomaly, cerebral palsy, amputations, and fractures if such impairment adversely affects a student's educational performance" (Kirk, Gallaher, Coleman & Anastasiow, 2011, p.4); cited from National Dissemination Center for Children with Disabilities [NICHCY], 2013).

This category can be divided into two main areas: (a) Neuro-motor impairments and (b) Muscular/Skeletal conditions. Due to increased medical technological discoveries we have reduced prevalence rates in some conditions such as Polio, and Spina Bifida due to vaccination/greater awareness and the proactive use of Folic Acid respectively. Muscular dystrophy also exists at lower prevalence levels while conditions such as Multiple Sclerosis are quite rare in children. We are still more likely to encounter individuals that live with Cerebral Palsy, Epilepsy, and Spina Bifida within our special education classrooms.

Neuromotor Impairments. These conditions arise from damage afflicted to the central nervous system (CNS), which includes both the brain and spinal cord. Depending of the severity of the damage we can see a variety of ability of movement. Limited movement by muscles and their control can adversely affect school participation, engagement and task completion. Individuals with these conditions may have minor to significant ability to control their movements through the use of specific devices which help them become more independent allowing for talking, walking and writing. In addition to what can be seen as somewhat obvious with PIs, it can be easy to miss the social and emotional toll that can result from an inability to communicate and build friendships; relationships and the ability to communicate freely. Being understood accurately (being able to say what one truly feels) is crucial (Coster & Haltiwanger, 2004). Seizure disorders also fall under this section, where treatment with medication has made an impact and improved functioning within schools and communities. Pascalicchio, de AraujoFilho, da Silva Noffs, Lin, S.F. Caboclo, Vidal-Dourado, Ferreira Guilhoto, and Yacubian (2007) conducted a study of 50 individuals aged ≥ 17 that had Juvenile Myoclonic Epilepsy (JME) with a control group who did not. A series of neuropsychological tests were conducted looking at executive functioning, cognitive function, memory, and vocabulary. In each section the focus group scored lower than that of the control; however what was important to note was the role schooling played in the attained scores. They felt that schooling provided a protective factor of sorts in that those who had ≥11 years of educational experience fared better in their testing than those individuals who had <11 years experience. In another study conducted by Papavasiliou, Mattheou, Bazigou, Kotsalis, and Paraskevoulakos, (2005) individuals with epilepsy performed significantly worse than the control group in spelling, reading aloud, and reading comprehension; they presented dyslexic-type errors; and frequently had below-average school performance.

The condition of CP falls under the supports of special education and although it is not a disease it has significant affect on ones quality of life. It can occur prior to, during and after birth. The presentation of symptoms varies across a spectrum where muscle coordination and body movement may present minimally or significantly. Majnemer, Shevell, Law, Birnbaum, Chilingaryan, Rosenbaum, and Poulin, (2008) conducted a study on individuals who live with CP within the school setting; the children with CP demonstrated limitations in their ability to participate in playground and recess activities. Evidence suggests that children with a variety of disabilities are involved in fewer activities than their peers, and that these activities tend to be home-based and less physically active, with fewer social engagements.

Limb deficiencies also feature under the muscular skeletal conditions. These can be congenital or as the result of HI or an accident. Modern technologies have paved the way toward the production of replacement limbs that function in a manner that allow for greater inclusion for that individual both within school and the community (Turnbull, Turnbull, and Meyer, 2010). Amongst the conditions that affect muscles and joints Juvenile Arthritis (JA) is relatively common, though it falls within the OHI catchment. Approximately 25% of students with JA receive special education services. Others under this label can and do receive services under a 504 plan. The symptoms revolve around immobility, pain, side effects of treatments and bone or joint damage.

Health Impairments

"Chronic or acute health problems such that the physiological capacity to function is significantly limited or impaired and that results in limited strength, vitality, or alertness, including a heightened alertness to environmental stimuli, resulting in limited alertness with respect to the educational environment" (Donovan & Cross, 2002; Kirk, Gallaher, Coleman & Anastasiow, 2011, p.4). "Chronic or acute health problems such as asthma, attention deficit disorder or attention deficit hyperactivity disorder, diabetes, epilepsy, a heart condition, hemophilia, lead poisoning, leukemia, nephritis, rheumatic fever, sickle cell anemia, and Tourette syndrome are all conditions which adversely affects a child's educational performance." (National Dissemination Center for Children with Disabilities, 2013, p.4)

Modern technology and advances have now set the stage for greater life quality and longevity. Students can move in and out of crises or fragile status and survive. Even though many are now stable enough to attend school, they require ongoing medical management. It is here the connection and collaborative efforts need to be manifested in a clear and sustainable way, to allow for quality of life, equal opportunity to all curriculum and success. The necessary supports must be detailed within the IEP and teachers should be knowledgeable and skilled at provision or reaching those who are proficient (Heller, Hsieh, & Rimmer, 2004; Heller & Tumlin, 2004). "Although the contingencies for the "worst case scenarios" must be arranged, in most cases the accommodations required for these students are not terribly dramatic" (Smith& Tyler, 2010, p.4).

The chronic illnesses/impairments can be divided into two groups (1) chronic disease and (2) infectious diseases (Smith & Tyler, 2010). When an individual lives with a chronic health condition the impact can be substantial; valuable time in terms of access to curriculum can be missed, the symptoms which may be ever present or intermittent can detract from the learning process and treatment comes with its own set of side effects that in their quest to impede the condition may exacerbate other parts of the body which affect the whole being and self esteem (e.g. Chemotherapy and the loss of hair; use of steroids and weight gain or mood change). Modern medicine has allowed for the improved life quality for those individuals that live with chronic health and/or physical impairments; so much so that previously the individual may not have been schooled within a public regular school but perhaps home-schooled or via a hospital based program (Smith & Tyler, 2010). Through the passing of IDEA mandates we now have the responsibility for not only physical inclusion but also its counterpart academic inclusion.

Infectious Diseases

These categories of chronic health or shorter more intense diseases breed a sense of unease and we have wavered in the policies surrounding disclosure to school populations, teachers and staff as a whole. As of 2002, a total of 9,300 American children were infected with Human Immune Deficiency Virus /Acquired Immune Deficiency Syndrome (Fauci, Touchette, & Folkers, (2005).

Conditions such as HIV have finally began to shed some of its stigma as we now have a greater depth of understanding of its complexities and process as well as the provision of a treatment protocol that sets the stage for a significant improvement in life expectancy and quality. Being HIV positive does not mean one has AIDS and in fact the Aids Related Complex (ARC) is the stepping-stone between the two. The transition can take years with current treatment or months without the appropriate care. It is fatal and the concerns (which are not unfounded due to the increased transferability factors) make it one condition schools have considerable anxiety around.

Science, Technology, Engineering and Mathematics (STEM)

Atkinson & Mayo (2010) reviewed the need for STEM approaches and those approaches which have prevailed, those that have failed and those that are being created to keep the U.S. as a contender amongst other countries focusing on STEM innovations (Augustine, 2005; Kuenzi, Matthews, & Mangan, 2006; Langen & Dekkers, 2005; U.S. Department of Education: National Center for Educational Statistics, 2009). The seriousness of America's future is detailed in terms of how many graduates within STEM it can produce. An approach was to increase access of STEM for all; which "ensures that all students get as much STEM as possible at each step in their education" (Kuenzi, Matthews, & Mangan, 2006) (p. 1)

Issues. Boosting the quality of STEM instruction [through Teacher preparation programs] (Kuenzi, Matthews, & Mangan, 2006), adapting the curriculum to strengthen its STEM content and building a thorough STEM standards approach is warranted. Trying at the same time to make STEM more exciting for students so that we build a pipeline into STEM fields is also needed. Atkins and Mayo (2010) argued the limitations of this approach: (1) only 5 percent of the workforce are scientists and engineers, thus the amount to invest in an 'access of STEM for all' would be considerable and perhaps irresponsible. The National Assessment of Educational Progress (NAEP) has assessed students since 1969 in the areas of math and science knowledge. They report their data on a scale denoted as basic, proficient and advanced. Data from 1990-2005 show an increase in students achieving proficiency at the 4th and 8th grade in science, however math scores remain basic. There is no data for engineering or technology subjects (Kuenzi, Matthews, & Mangan, 2006); (2) The 'STEM for All' approach removes from our educational system the respect for individual desires to be active contributors, and learners in their own educational journey. The imposition of increased STEM curricula and standardized testing around these areas may not make students more likely to become STEM workers. They cite that over 80 percent of STEM jobs are in engineering and information technology, yet engineering and computer science are barely present within the K-12 system curriculum. Making these arenas accessible is key to building our capacity in the workforce.

Suggestions. 'All STEM for some' where the onus is on the recruitment of those individuals who have a keen interest in STEM with a emphasis on underrepresented groups. This allows a more focused approach to providing STEM curricula access and increases the chances of successful experiences and a culmination of capacity for the STEM U.S. innovative economy. One might argue in this case that some individuals may not know that they are interested in until further participation in more dynamic curricular experiences. A summary of the 'Refueling the U.S. Innovation Economy: Fresh Approaches to Science, Technology, Engineering and Mathematics Education looks at (1) reasons for the shortage of practitioners, (2) limitations of current STEM approaches, (3) shifting from teaching STEM facts to teaching STEM skills, (4) building inter-disciplinary programs and curricular frameworks, and (5) matching more cohesively STEM graduates and the jobs that are available. These suggestions are not just a literature review but foresights into policy change and adaptation.

Setting the Stage. Another area of considerable affect is the impact that teachers have in the area of STEM. NAEP reviewed teacher quality over the past 20 years and noted that having a major in this area significantly improves outcomes for students however this is an area that is lacking in the practitioner field. Most recent surveys of teachers amounting to 1.4

million teachers from 1999-2000 showed that 13.7 percent were math majors and/or teaching math and 11.4 percent were science majors and/or teaching science. These numbers speak for themselves.

These data may provide a clue as to how we prepare teachers looking at the programs that have the capacity to provide school systems, and what is required to produce students that are motivated and proficient to move into STEM fields (Atkinson & Mayo , 2010; Kuenzi, Matthews, & Mangan, 2006).

The National Academy of Science (NAS), the National Academy of Engineering (NAE) and the Institute of Medicine (IM) issued a joint report entitled: *Rising Above the Gathering Storm: Energizing and Employing America for a Brighter Economic Future.* This narrative focused on increasing the robustness of the STEM pipeline from elementary through postsecondary education (NAS, 2005).

Their suggestions mirrored those mentioned above strengthening teaching force, increased investment into STEM programs, increasing postsecondary pools of students pursuing STEM degrees. The National Science Foundation (NSF) broadened what was included in STEM by adding natural sciences, social and behavioral sciences and computer information sciences.

The statistics of who enters into the STEM fields are of interest, we see a greater proportion of men to women 33 percent to 14 percent; ethnicity plays a role, 47 percent of Asian/Pacific Islander compared to all other ethnic groups who had no measurable difference; dependent and younger students outweighed older (> 24 years old), as did foreign born students compared to the U.S. born counterparts. K-12 indicators were telling, showing students who took pre-calculus, trigonometry, and had a higher than 'b" grade point average were expected to attain a graduate degree in STEM while those without those experiences were less likely (Chen, 2009).

A historical piece is also noted in families which have a high school education background (parents) and those that earn within the to 25 percent income bracket of the population are more likely to engage in a STEM career path.

This leads us to a very important question about students identified as having a disability and experiencing the special educational journey that is currently repeated in schools across the U.S. The statistics mentioned above show an exclusionary pattern of sorts for students within both public and private schools. Special education not being mentioned anywhere (see above citations).

When we hear 'STEM for ALL' we understand that 'ALL' leaves no one out however we have a conundrum when we see the listed indicators and perhaps predictive statistics of those students who are currently moving into the STEM career pathway. This implies a particular series of gates must be opened in order to provide access.

Harris, Bransford, and Brophy (2002) reviewed the teaching and learning frameworks of biomedical engineering and noted that that best practice involved learning environments that are learner centered, knowledge centered, assessment and community centered. These structures are reminiscent of frameworks, which are noted as best practices for special education (Turnbull, Turnbull & Wehmeyer, 2010). Current research has shown that the traditional methodologies of teaching STEM has been sufficient only for the few that can maintain the interest of such rigorous scientific and mathematical journey's through their schooling.

Integration

The use of an integrative approach represents a "robust new professional development opportunity" (Mayo, 2009; Sanders, 2009, p.22). This professional development must be sustained over time and site based. Linking the inherent connections between science, math, and technology. Sanders (2009) also makes clear that STEM education and STEM content are not the same thing; This understanding impacts our interventions. Ostler, (2012) stated that the blanket manner in which we approach STEM intervention can be the cause of inconsistent progress. He stated that STEM should look different at different stages and perspectives. Where teacher preparation is concerned we need to have solid content knowledge but also a deeper integrative understanding in order to facilitate integrative approaches. The issue of assessment is also brought up where he reflects on current standardized assessments as being inappropriate to measure true integrative STEM knowledge. Berry, Chalmers, and Chandra (2012) also looked at the pedagogical approaches to the STEM subjects and the manner in which subjects are taught and how despite departments being co-located within schools the teaching of each subject can be quite different which is a barrier to integrative STEM education. The essence of integrative STEM education is founded in the framework of constructivism; looking at learning as a constructive not receptive process; the beliefs and motivations of students are central to cognition; social interaction is fundamental to cognitive development, and the grounding of knowledge within context crucial (Bruning, Schraw, Norby & Ronning, 2004). If integration is to be implemented with intent then we must see it through K-12 curriculum. Currently we see a paucity of technology and engineering content within the K-5 curriculum (NCES, 2008).

Helal, Mokhtari, & Abdulrazak (2008) cite Irving P Schloss who makes a statement regarding the national science and technology policy organization and priorities act (NSTPOPA) of 1976; where the omission of disabilities or the "handicapped' and how those individuals can be served is a serious oversight by the organization. They go further to analyze policy regarding technology and disability; here they unpack the NSTPOPA and decipher where support for those living with disabilities is embedded. This document however looked at post schooling age range and how the general policies support individuals such as Americans with Disabilities Act (ADA) and how it covers employment and environmental access incorporating universal design (UD). They speak on technology specific laws such as the Decoder Circuitry Act of 1990 (DCA), which looks at smart technology. These universal approaches cover what they term "consumers for independence" (p.63). They examined allotted budget and deem that due to low funding allowance the associated research is lacking and overdue.

Special Education and STEM: Its Importance

"Special education research, because of its complexity, may be the hardest of the hardest-to-do. One feature of special education research that makes it more complex is the variability of the participants." (Odom, Brantlinger, Gersten, Horner, Thomson & Harris, 2005, p.137). Gulnick (1999) cited in Greene, (2013) spoke on the specificity of research within special education; how we as a field have a number of contexts to consider and interventions, making for a complicated outcome or generalizable piece of data. Therefore it is important to look at the issues and opportunities that are available for students with both physical and health impairments to access the STEM curricula.

The goal being to increase their opportunity for both engagement in those fields which resultantly will set them up to bridge the gap seen in the data of who are over and under-represented in these fields. This is fulfilling the legal mandate in special education for transition to adulthood, employment and success in school (Artiles & Trent, 1994; Turnbull, Turnbull, & Meyer, 2010).

Burgstahler (1994) posed this problem in her study "Individuals with disabilities are under-represented in science, engineering, and mathematics educational programs and professions.

Causes of this problem can be found in three areas: (1) preparation of students with disabilities; (2) access to facilities, programs, and equipment; (3) and acceptance by educators, employers, and co-workers." (p.1). It has been nearly two decades and we are still suffering from a scarcity of answers and solutions to the issue. She goes on to state as support for her endeavor that the accessibility to higher education through appropriate high school education for those individuals with disabilities (physical and health) enhances employability and "vocational success" (p.1).

She added that though increasing numbers of individuals with disabilities were attending post-secondary institutions, few engaged in programs that were science or engineering based, and when they did the attrition rate was high. Even if completion of a program occurred the employment rate of the individual was lower in ratio to their typically developing counterpart. A very interesting note was that in the group of individuals with PI the employment rate within the field of sciences was higher than that group as a whole in other fields indicating a suitability and appropriateness for the successful participation in the math and science fields.

Strategies to Enhance Success at the K-12, Transition and Post-Secondary Level

Ostler (2012) looked at traditional and innovative ways to build STEM areas within K-12 education. He states that the use of practical and traditional academic facts can lead to the manifestation of STEM products. He also contends with the issues around preparing STEM educators as a conduit to motivating and creating success in secondary as well as post secondary experiences.

Listed are interventions, which come in a generalized form to support the gaps in our current practice: (1) All individuals have *access* to *technology* that provides positive academic and career outcomes; (2) People with disabilities learn to *use technologies* in ways that contribute to positive post-secondary academic and career outcomes and self-determined lives; and (3) There is a *seamless transition* of availability for technology for all people with disabilities as they move from K-12 to post secondary to career environments (Burgstahler, 2003; 2011).

For conditions where absence and the ability to focus or retain information for longer than short term memory, allows for some clear strategies which emphasize routine, clarity of information, consistent communication with teacher and peers and use of technology to support processing and memory. This is where the use of technology plays a significant role where teachers and classrooms can share information while a child is out of the class. The objectives of the class and subject need to remain challenging, motivating and engaging. Assessment and progress should be communicated and this further supports a sense of being a part of the community. Teachers can set the stage for 'errorless learning' which research states enables information that is learned to be retained for longer periods (Turnbull, Turnbull, & Meyer, 2010). Further detail and utilization of UD understanding and framework to this context is envisioned by Bowe (2000) and displayed in Table 1.

Table 1. Utilization of Universal Design for Learning

Activity	Strategy
Maximize independence in academic and employment tasks	Utilize a hands free keyboard and mouse
Participate in classroom discussions	Use of communication devices (assistive technology);
Gain access to peers, mentors, and role models	Supported internet community;
Self-advocate for internship	Use of TTY and relay service accommodations and supports
Gain access to full range of educational options	Use of universal design for PI, HI or visual impairments
Participate in experiences not otherwise possible	Engage in activities through virtual reality and simulations with technology.
Secure high levels of independent living	Contact caregivers through a respectful independent distance given, with the technology to navigate living activities such as cooking, TV, internet etc…
Prepare for transitions to college careers	Clear and consistent transition supports that start in high school with support provided from university level disability support services. Communication between support services and internship personnel and special field placements.
Succeed in work-based learning experiences	Use of technology (UD) to complete aspects of tasks such as experiments, creation of models and math proofs.
Work side-by-side with peers	Build a reverse inclusion model where typically developing individuals may be in the minority and so all skills are utilized and warranted.
Mastery of academic tasks previously un attainable	Use of appropriate, technologies, software and treatments that bridges the gap is skill or ability (physical, processing or health related).
Enter high tech career fields	The fielding and support of early interests. The use of UD, early supports and advocacy fulfill a motivating and make clear the 'high expectations'.
Participating in recreational activities	Bridge the social and recreational gap into community activities and programs where universal design and expectations are normalized so the goal is for *all* to become a part and contribute.

CONCLUSION

Improving education in STEM in U.S. public schools has been at the forefront of educational issues and a national priority for over forty years. "Initiatives like *Changing the Equation*, a part of the *Educate to Innovate Campaign* focuses on (1) allowing more students to engage in robotics competitions, (2) improving professional development for math and science teachers, (3) increasing the number of students that take and pass rigorous Advanced Placement (AP) math and science courses, (4) increasing the number of teachers who enter the profession with a STEM undergraduate degree and (5) providing new opportunities to traditionally underrepresented students and underserved communities" (Green & Turton, in press)

A substantial shift in our programmatic structures needs to occur to facilitate an equal opportunity access for our students of all diverse needs. When this is coupled with the considerations and descriptions of current understandings regarding physical and health conditions which have a long-term effects on a child's learning and participation one can find themselves in a quandary. Through this we can open a new door for a new type of student and student experience. We can harness an untapped source, which will benefit all. There is a need to recognize the existing landscapes of STEM, PI/HI, current educational options and suggested directions in which to move forward.

REFERENCES

Anderson, C.L., & Petch-Hogan, B. (2001). The impact of technology use in special education field experience on preservice teachers' perceived technology expertise. *Journal of Special Education Technology, 16*(3), 27-39.

Artiles, A. J., & Trent, S. C. (1994). Overrepresentation of Minority Students in Special Education A Continuing Debate. *The Journal of Special Education*, *27*(4), 410-437.

Atkinson, R., & Mayo, M. (2010). Refueling the US Innovation Economy: Fresh Approaches to Science, Technology, Engineering and Mathematics (STEM) Education. Executive Summary. *Information Technology and Innovation Foundation.*

Augustine, N. R. (2005). Rising above the gathering storm: Energizing and employing America for a brighter economic future. *Retrieved February 14th, 2014.*

Berry, M. R., Chalmers, C., & Chandra, V. (2012). STEM futures and practice, can we teach STEM in a more meaningful and integrated way?

Bowe, F.G. (2000). Universal design in education. Westport, CT: Bergin & Garvey.

Brown, R. T. (Ed.). (2003). *Handbook of pediatric psychology in school settings.* Routledge.

Brown, R. T., Buchanan, I., Doepke, K., Eckman, J. R., Baldwin, K., Goonan, B., & Schoenherr, S. (1993). Cognitive and academic functioning in children with sickle-cell disease. *Journal of Clinical Child Psychology*, *22*(2), 207-218.

Bruning, R. H., Schraw, J. G., Norby, M. M., & Ronning, R. R. (2004). Cognitive psychology and instruction. Columbus, OH: Pearson.

Burgstahler, S. (1994). Increasing the representation of people with disabilities in science, engineering, and mathematics. *Information Technology and Disability*, *1*(4).

Burgstahler, S. (2003). The role of technology in preparing youth with disabilities for ostsecondary education and employment. *Journal of Special Education Technology, 18*(4), 7-20.

Burgstahler, S., Comden, D., Lee, S. M., Arnold, A., & Brown, K. (2011). Computer and cell phone access for individuals with mobility impairments: An overview and case studies. *Neuro Rehabilitation, 28*(3), 183-197.

Carter, E. W., Cushing, L. S., Clark, N. M., & Kennedy, C. H. (2005). Effects of peer support interventions on students' access to the general curriculum and social interactions. *Research and Practice for Persons with Severe Disabilities, 30,* 15-25.

Chapa, J., & De La Rosa, B. (2006). The Problematic Pipeline Demographic Trends and Latino Participation in Graduate Science, Technology, Engineering, and Mathematics Programs. *Journal of Hispanic Higher Education, 5*(3), 203-221.

Chen, X. (2009). Students Who Study Science, Technology, Engineering, and Mathematics (STEM) in Postsecondary Education. Stats in Brief. NCES 2009-161. *National Center for Education Statistics.*

Coster, W. J., & Haltiwanger, J. T. (2004). Social—behavioral skills of elementary students with physical disabilities included in general education classrooms. *Remedial and Special Education, 25*(2), 95-103.

Donovan, M. S., & Cross, C. T. (Eds.). (2002). *Minority students in special and gifted education in today's schools (6th ed.).* Columbus, Ohio: Merrill Prentice Hall. *education.* National Academies Press.

Fagot-Campagna, A., Pettitt, D. J., Engelgau, M. M., Burrows, N. R., Geiss, L. S., Valdez, R., & Venkat Narayan, K. M. (2000). Type 2 diabetes among North adolescents: An epidemiologic health perspective. *The Journal of pediatrics, 136*(5), 664-672.

Faraone, S. V., Sergeant, J., Gillberg, C., & Biederman, J. (2003). The worldwide prevalence of ADHD: Is it an American condition? *World Psychiatry, 2*(2), 104.

Hardman, M. L., & Clark, C. (2006). Promoting friendship through Best Buddies: A national survey of college program participants. *Journal Information, 44*(1).

Fauci, A. S., Touchette, N. A., & Folkers, G. K. (2005). Emerging infectious diseases: a 10-year perspective from the National Institute of Allergy and Infectious Diseases. *The International Journal of Risk and Safety in Medicine, 17*(3), 157-167.

Goldstein, J., Plioplys, S., Zelko, F., Mass, S., Corns, C., Blaufuss, R., & Nordli, D. (2004). Multidisciplinary approach to childhood epilepsy: exploring the scientific rationale and practical aspects of implementation. *Journal of child neurology, 19*(5), 362-378.

Greene, S. (2013). *Race, Community, and Urban Schools: Partnering with African American Families.* Teachers College Press. NY.
Hardman, M. L., & Clark, C. (2006). Promoting friendship through Best Buddies: A national survey of college program participants. *Journal Information, 44*(1).
Harris, T. R., Bransford, J. D., & Brophy, S. P. (2002). Roles for learning sciences and learning technologies in biomedical engineering education: a review of recent advances. *Annual Review of Biomedical Engineering, 4*(1), 29-48.

Hasselbring, T. S., & Glaser, C. H. W. (2000). Use of computer technology to help students with special needs. *The Future of Children,* 102-122.

Helal, A.A., Mokhtari, M., & Abdulrazak, B. (Eds.). (2008). *The engineering handbook of smart technology for aging, disability, and independence* (Vol. 10, p. 9780470379424). Hoboken, NJ: Wiley.

Heller, T., Hsieh, K., & Rimmer, J. (2004). Attitudinal and psychosocial outcomes of a fitness and health education program on adults with Down syndrome. American Journal on Mental Retardation, 109, 175–185.

Heller, K. W., & Tumlin, J. (2004). Using expanded individualized health care plans to assist teachers of students with complex health care needs. *The Journal of School Nursing, 20*(3), 150-160.

Holland, S. (2014). The pen, phone and iPad: Obama's tech push in U.S. schools. Retrieved from http://www.reuters.com/article/2014/02/04/us-usa-obama-education-idUSBREA130J520140204

Holmes, C. S., Cant, M. C., Fox, M. A., Lampert, N. L., & Greer, T. (1999). Disease and demographic risk factors for disrupted cognitive functioning in children with insulin-dependent diabetes mellitus (IDDM). *School Psychology Review, 28*, 215-227.

Huberty, T. J., Austin, J. K., Huster, G. A., & Dunn, D. W. (2000). Relations of change in condition severity and school self-concept to change in achievement-related behavior in children with asthma or epilepsy. *Journal of School Psychology, 38*(3), 259-276.

Kirk, S., Gallagher, J., Coleman, M., & Anastasiow, N. (2011). Educating Exceptional Children. Cengage, Wadsworth, USA.

Konopasek, D. E., & Forness, S. R. (2004). Psychopharmacology in the treatment of emotional and behavioral disorders. *Handbook of research in emotional and behavioral disorders*, 352-368.

Kuenzi, J. J., Matthews, C. M., & Mangan, B. F. (2006). Science, technology, engineering, and mathematics (STEM) education issues and legislative options. Library of Congress Washington DC congressional research service.

Langen, A. V., & Dekkers, H. (2005). Cross national differences in participating in tertiary science, technology, engineering and mathematics education. *Comparative Education, 41*(3), 329-350.

Majnemer, A., Shevell, M., Law, M., Birnbaum, R., Chilingaryan, G., Rosenbaum, P., & Poulin, C. (2008). Participation and enjoyment of leisure activities in school aged children with cerebral palsy. *Developmental Medicine & Child Neurology, 50*(10), 751-758.

Mayo, M. J. (2009). Video games: A route to large-scale STEM education? *Science, 323*(5910), 79-82.

McDonnell, J. J., Hardman, M. L., & McDonnell, A. P. (2003). *An introduction to persons with moderate and severe disabilities: Educational and social issues*. Allyn and Bacon.

McPherson, M., Arango, P., Fox, H., Lauver, C., McManus, M., Newacheck, P. W., & Strickland, B. (1998). A new definition of children with special health care needs. *Pediatrics, 102*(1), 137-139.

Merikangas, K. R., He, J. P., Burstein, M., Swanson, S. A., Avenevoli, S., Cui, L., & Swendsen, J. (2010). Lifetime prevalence of mental disorders in US adolescents: results from the National Comorbidity Survey Replication–Adolescent Supplement (NCS-A). *Journal of the American Academy of Child & Adolescent Psychiatry, 49*(10), 980-989.

National Children's Study Interagency Coordinating Committee. (2003). The National Children's Study of environmental effects on child health and development. *Environ Health Perspect.* 111, 642–646.

National Dissemination Center for Children with Disabilities. Retrieved from http://www.nichcy.org/.

Newacheck, P. W., & Halfon, N. (1998). Prevalence and impact of disabling chronic conditions in childhood. *American Journal of Public Health*, *88*(4), 610-617.

Odom, S. L., Brantlinger, E., Gersten, R., Horner, R. H., Thompson, B., & Harris, K. R. (2005). Research in special education: Scientific methods and evidence-based practices. *Exceptional Children*, *71*(2), 137-148.

Ostler, E. (2012). 21st Century STEM Education: A Tactical Model for Long-Range Success. *International Journal of Applied*, *2*(1).

Papavasiliou, A., Mattheou, D., Bazigou, H., Kotsalis, C., & Paraskevoulakos, E. (2005). Written language skills in children with benign childhood epilepsy with centrotemporal spikes. *Epilepsy & Behavior*, *6*(1), 50-58.

Pascalicchio, T. F., de AraujoFilho, G. M., da Silva Noffs, M. H., Lin, K., Caboclo, L. O. S., Vidal-Dourado, M., &Yacubian, E. M. T. (2007). Neuropsychological profile of patients with juvenile myoclonic epilepsy: a controlled study of 50 patients. *Epilepsy & Behavior*, *10*(2), 263-267.

Pinhas-Hamiel, O., & Zeitler, P. (2005). The global spread of type 2 diabetes mellitus in children and adolescents. *The Journal of pediatrics*, *146*(5), 693-700.

Pinhas-Hamiel, O., Dolan, L. M., Daniels, S. R., Standiford, D., Khoury, P. R., & Zeitler, P. (1996). Increased incidence of non-insulin-dependent diabetes mellitus among adolescents. *The Journal of pediatrics*, *128*(5), 608-615.

Pottie, C., & Sumarah, J. (2004). Friendships between persons with and without developmental disabilities. *Journal Information*, *42*(1).

Quinn, M. M., Rutherford, R. B., & Leone, P. E. (2001). *Students with disabilities in correctional facilities*. ERIC Clearinghouse on Disabilities and Gifted Education.

Roco, M. C., & Bainbridge, W. S. (2002). Converging technologies for improving human performance: Integrating from the nanoscale. *Journal of Nanoparticle research*, *4*(4), 281-295.

Reichman, N. E., Corman, H., & Noonan, K. (2008). Impact of child disability on the family. *Maternal and child health journal*, *12*(6), 679-683.

Rovet, J. F., Ehrlich, R. M., & Hoppe, M. (1988). Specific intellectual deficits in children with early onset diabetes mellitus. *Child development*, 226-234.

Sanders, M. (2009). STEM, STEM education, stemmania. *The Technology Teacher*, *68*(4), 20-26.

Smith, D. D., & Tyler, N. C. (2010). *Introduction to special education: Making a difference*. Merrill/Pearson Education. Steinberg, M. H. (1999).Management of sickle cell disease. *New England Journal of Medicine*, *340*(13), 1021-1030.

Sturge, C., Garralda, M. E., Boissin, M., Dore, C. J., & Woo, P. (1997). School attendance and juvenile chronic arthritis. *Rheumatology*, *36*(11), 1218-1223.Summers, J. A., Hoffman, L, Marquis, J., Turnbull, A., & Poston, D. (2005).

Relationship between parent satisfaction regarding partnerships with professionals and age of child. *Topics in Early Childhood Special Education*, *25*(1), 48-58.

U.S Department of Education. (2008). Report to Congress on the Implementation of the Individuals with Disabilities Act, Vol 1. Retrieved from: http://nces.ed.gov/pubs2007/2007024.pdf

Wodrich, D. L., & Schmitt, A. J. (2003). Pediatric topics in the school psychology literature: Publications since 1981. *Journal of School Psychology*, *41*(2), 131-141.

Wodrich, D. L., & Spencer, M. L. (2007). The other health impairment category and health-based classroom accommodations: School psychologists' perceptions and practices. *Journal of Applied School Psychology*, *24*(1), 109-125.

Wodrich, D. L., Pfeiffer, S. I., & Landau, S. (2008). Contemplating the new DSM-V: Considerations from psychologists who work with school children. *Professional Psychology: Research and Practice*, *39*(6), 626.

In: S.T.E.M. Education
Editor: Satasha L. Green

ISBN: 978-1-63117-806-1

Chapter 9

WORKING WITH LEARNERS WITH HEARING LOSS IN STEM

*C. Jonah Eleweke**
College of Liberal Arts and Sciences
World Languages and Literatures
Portland State University
Portland, Oregon

ABSTRACT

Learners with hearing loss can acquire science, technology, engineering, and mathematics (STEM) knowledge and skills. There is evidence in history that several important inventions in STEM were made by individuals with hearing loss. The vast majority of young people with hearing loss come from families with no previous experience of raising children with hearing loss. It is argued that these families should be provided unbiased and clear information regarding all aspects of the development of children with hearing loss. With such information, they can make informed choices that will positively impact their involvement in early intervention programs for their children. Thus they will begin to lay the foundation for the development of STEM knowledge and skills in their children by being able to consistently communicate with them. The implications of utilizing educational materials and programs with rich visual effects such as videos, animations, 3-Ds, and subtitles; having effective teachers who plan and address the unique learning needs of learners with hearing loss; and supporting learners with hearing loss to be actively involved in their own learning process to facilitate the development of STEM knowledge and skills are discussed.

INTRODUCTION

In this chapter, important factors for understanding and enhancing the educational experiences, and outcomes for learners who have hearing loss in science, technology,

* jeleweke@pdx.edu

engineering, and mathematics (STEM) are highlighted. Hearing loss is used in this chapter as a broad generic term for any hearing dysfunction that precludes successful processing of all linguistic information through the ears regardless of severity, etiology, and age of onset. Nonetheless, the different degrees of hearing loss: mild, moderate, severe, and profound clearly suggest that children who have hearing loss have different communication and educational needs (see for example, Gomez, Piehota, & Dischner, 2012; McKay, Gravel, & Tharpe, 2008) for a review.

In the United States, Canada, United Kingdom, and other Western countries, there have been concerted efforts toward developing strategies for addressing STEM and other educational needs of individuals with hearing loss over the past decades (Spencer & Marschark, 2010). This has been due to increased understanding of people with hearing loss, their communication methods, and ongoing changes and progress in pedagogy, psychology, psycholinguistics, and technology. Evidence suggests that more progress has been made in the education of children with hearing loss in the United States during the last 30 years than in the previous 300 years (Marschark, Lang, & Albertini, 2002).

Although history is replete with amazing feats of many famous individuals with hearing loss in STEM, for example, Thomas Edison's light bulb and thousands of other inventions; Sir John Warcup Cornfield, winner of the Nobel Prize in 1975 for his work in Chemistry and cholesterol; Robert Weitbretch, inventor of the special telephone for deaf people (teletypewriter or TTY, for short); Annie Jump Ganon, in astronomy; to mention just a few (Deaf Scientist Corner, 2013); current evidence indicates that majority of learners with hearing loss have lower achievements in STEM compared to their hearing peers (Blatto-Vallee, Kelly, Porter, & Fonzi, 2007; Vosganoff, Paatsch, & Toe, 2011).

Consequently, new options and perspectives that could lead to achieving better outcomes for learners with hearing loss in STEM should be continually provided to families and their children with hearing loss, their teachers, and other professionals in the field. Controversies regarding the best strategies for instructions and services to achieve this goal, especially in terms of methods of communication, have continued to be prevalent in the field (Mukaria & Eleweke, 2010). Nonetheless, it is argued that learners with hearing loss can achieve success in STEM if such controversies are set aside, and efforts are focused on providing appropriate services that will foster their learning and cognitive development immediately following the diagnosis of hearing loss. Given that the majority of children with hearing loss in this country and other Western countries now live at home and are in inclusion educational programs due to numerous legislative mandates (Hyde & Power, 2003; Marschark, Sapere, Convertino, & Pelz, 2008) programs and services to enhance their educational development and subsequent success in STEM must begin immediately after the diagnosis of hearing loss. To effectively provide meaningful foundation for the success of children with hearing loss in acquiring STEM knowledge and skills, the following are important considerations: (1) supporting families, (2) appropriate and effective early intervention programs, (3) use of relevant technology and software, and (4) effective teaching strategies and study habits.

SUPPORTING FAMILIES TO PROVIDE STEM DEVELOPMENT SUCCESS

Majority of children with hearing loss in the United States, Canada, United Kingdom and most Western countries live at homes with their families due to legislative mandates supporting inclusion of learners with special needs in regular education programs in these countries (see for example, Marschark, Sapere, Convertino & Pelz, 2008). Consequently, the foundation for the success of these children in all aspects of development including STEM must be laid at homes by supporting families. Evidence indicates that the vast majority of children with hearing loss, more than 90%, come from hearing families with no previous experience of hearing loss (Mitchell & Karchmer, 2004). With no previous experience of raising children with hearing loss, the parents could experience various devastating psychological reactions when the hearing loss was diagnosed and disclosed to them (Borum, 2012; Jackson, 2011). Without timely support, some parents could remain in denial and shopping around for a 'cure' for too long (Scheetz, 2012). There is a real danger that the child's developmental needs could be neglected if the adverse psychological reactions parents may experience by the diagnosis of hearing loss persisted for too long. These parents, therefore, require timely support and education regarding childhood hearing loss and its implications for the child's development. They need to know about the various communication and educational options for children with hearing loss (Marschark, 2007).

For parents to be effective in meeting the developmental needs of their children with hearing loss and ensure their success in all aspects of development including STEM, they should be empowered to take the initiative in seeking out programs and services that would foster the development of their children. To achieve this goal, families of children with hearing loss require clear and unbiased information on all the issues that are essential for the communication, educational, social-emotional and other aspects of the development of their children with hearing loss (Jamison, Zaidman-Zait, & Poon, 2011). Although the importance of providing such information is acknowledged in the literature, it seems to the case that in many instances, parents may not receive such information (Eleweke, Bays, Gilbert, & Austin, 2008). The failure to provide clear and unbiased information has adverse consequences. For instance, a family's ability to adapt to the child's hearing loss may be hindered by the lack of information about available services. Evidence clearly indicates that many parents are still not receiving adequate information and support during the critical early development years, and many remain uninformed and unable to assist their children at home. Lacking information, many parents are unable to make appropriate choices about the communication and educational needs of their children (Davila, 2004).

Providing unbiased and clear information is the critical modality of supporting families with young children who have hearing loss to enable them to lay the foundation for the development of STEM knowledge and skills in young children with hearing loss. When unbiased and clear information is provided, they will be able to make informed choices that will address the STEM knowledge and skills and other developmental needs of their children. However, concerns remain about the nature of the information professionals provide to families. Evidence indicates that many parents are aware that they are not getting all the information they need, and that some of the information they receive is patently biased (Marschark, 2007).

Nonetheless, evidence is consistent that the provision of unbiased and clear information and support are critically important to enable families to participate actively in the development and education of their children with hearing loss (Eleweke, et al., 2008; Jamison, Zaidman-Zait, & Poon, 2011). The information and support provided to families could make the biggest differences in their lives by empowering them to take the initiative to seek services that would help them and their children (Turnbull, Turnbull, Shank, & Smith, 2004). This demands that the family-centered approach which calls for collective empowerment rather than a more traditional relationship where the professional had power over the family should be utilized in sharing information and providing support to parents of children with hearing loss. With this approach, parents and family members must become essential partners in the decision-making process. Consequently, teachers and other service providers should consider the family constellation as the appropriate focal point of professional attention in providing information about developmental needs of children with hearing loss and supporting their families to follow through.

Using family-centered approaches will assist parents of children with hearing loss to acquire the information and supports they need in order to foster the STEM and other aspects of their children's development. For instance, a family's ability to adapt to the child's hearing loss may be facilitated by the provision of relevant information about available services. Equipped with appropriate information, support, and knowledge, many parents of children with hearing loss will be able to make appropriate choices about the communication and educational needs of their children, which will in turn facilitate the development of STEM knowledge and skills in the children. Evidence from the Colorado Home Intervention Program (CHIP) and studies in the United Kingdom indicated that families with young children with hearing loss could benefit immensely from programs provided within their homes (Yoshinaga-Itano, 2004). Parents could be visited by professionals such as teachers, speech and language pathologists, audiologists, early childhood special educators, bilingual educators, psychologists, and social workers. These professionals provide information about counseling, developmental assessment, communication, and educational options. Targets should be established for timing of initial visit by professionals to the family of a newly diagnosed child with hearing loss. The visits should be carefully planned and conducted with the goal of giving unbiased information and empowering parents so that they can make informed choices.

It is critical that professionals give contact information about local and national organizations concerned with hearing loss to the parents. These professionals, especially those with hearing loss, should be included in the team that makes the initial visit to families since they have first-hand experience of living with a hearing loss. Consequently, apart from providing information and mentoring families, they could give hope and inspiration to families and their children. Additionally, it is imperative that parents with newly diagnosed children have the opportunity of meeting and chatting with other parents who have been or are currently working their way through similar issues. These parents provide mutual and social supports to each other to cope better in meeting the challenges of raising a child with a hearing loss. Such parents manifest better behavioral interactions and greater sensitivity to their children's communication and other needs (Turnbull & Turnbull, 2006).

EFFECTIVE EARLY INTERVENTION PROGRAMS SUPPORTING STEM KNOWLEDGE AND SKILLS DEVELOPMENT

Given that the majority of parents of children with hearing loss are hearing people with no previous experience of raising children with hearing loss, it becomes imperative that following information and supports provided to the families to deal with their adverse reactions to the diagnosis of the hearing loss and obtaining understanding of what raising children with hearing loss entails, effective early intervention measures should be established to foster the development of STEM knowledge and other skills in children with hearing loss. Effective early intervention services could greatly enrich the conceptual store of children with hearing loss. Through such programs, communication barriers posed by hearing loss could be dealt with as parents and family members are supported to learn how best to communicate with their children (Fulcher, Purcell, Baker, & Munro, 2012).

Timely initiation of early intervention services could prevent or greatly reduce the communication barriers posed by hearing loss (Holzinger, Fellinger, & Beitel, 2011). These services are necessary because of their focus on language development, parent–child communication, social skills, educational, cognitive and other aspects of development, and support for the utilization of any residual hearing the children with hearing loss may possess. All these areas are important for the development of STEM knowledge and skills in children with hearing loss. It has been argued that there is evidence of strong relationship between language skills, especially reading and writing, and performance on STEM tasks and assessments (Vosganoff, Paatsch, & Toe, 2011). Consequently, appropriate early intervention programs may provide parents with strategies for enhancing their children's language and educational development through sign language instruction, speech training skills, or both depending on the particular program (Kasai, Fukushima, Omori, Sugaya, & Ojima, 2012). The goal of early intervention must be to foster effective parent–child communication starting soon after diagnosis of hearing loss. This kind of communication is the best single predictor of success in STEM knowledge and skills and virtually all other areas of development of children with hearing loss, (Marschark, 2007). Clearly, effective interactions, with the most appropriate means of communication, between caregivers and the child will provide opportunities for involvement in most normal activities of childhood in both social and academic areas, which foster the development of STEM knowledge and skills.

Thus, an important goal of early intervention should be to provide children with hearing loss with the opportunity to acquire language in the appropriate modality (sign or spoken) by having accessible and competent language models. Consistent means of communication, in the appropriate modality that is congruent for the child, will create opportunities for positive social interactions critical for the social-emotional development and STEM educational success of all children including those with hearing loss. Evidence indicates that children with hearing loss who are identified early and participate in early intervention programs achieve significantly better language, speech, social, and emotional developments than those who are not identified early and who receive intervention at a later time (Fulcher, et al., 2012; Holzinger, et al., 2011). To provide effective early intervention for children and their families, the program should not be influenced by the controversy regarding the ***best*** communication approach that has characterized the field of education of children with hearing loss for decades. There seems to be a common agreement with Marschark (2007) that there is no single correct answer about the *best* or superior method of communication for children with hearing loss. These children are different and therefore their communication and other developmental needs will be different. Consequently, no one method can be applied to all or even most of them. The important thing is to establish, as early as possible, an effective mode of parent–child communication, one that the child can access fully (Calderon, 2000). This is the foundation for successful social, emotional, and academic, STEM knowledge and skills, and other aspects of development of children with hearing loss. Through objective assessment, open-mindedness, and willingness to be flexible, the mode of communication that would be appropriate for each child could be found.

Fostering STEM Knowledge and Skills Using Information Communication Technology and Software

Over the decades there has been tremendous developments in the field of information and communication technology (ICT) that provide great benefits to learners with or without special needs. For many learners without special needs, learning materials on the internet, world wide web (WWW), CDs/DVDs, flash-drives, and so on support their independent learning. Using these resources, they can become self-directed learners. They can use these resources to learn in the school and at home. Unfortunately, the same cannot be said for learners who have hearing loss. Given the poor literacy skills of the majority of these learners, (Debevc & Peljhan, 2004; Vosganoff, Paatsch, & Toe, 2011), these ICT resources must be in forms rich with visual reinforcements. It has been argued that learners with hearing loss must be considered as visual learners (Moores, 2010). Consequently, educational materials and resources meant to foster STEM knowledge and skills that are based in print or spoken English or other national languages may be of very little benefit to the majority of learners with hearing loss who have poor skills with English literacy (Coppens, Tellings, Schreuder, & Verhoeven, 2013). Nonetheless, evidence suggests that ICT resources rich with visual elements can enhance the teaching and learning of STEM materials in learners with hearing loss (Debevc & Peljhan, 2004). According to these authors, the inclusion of visual or multi-media elements such as animation, videos, and subtitles in ICT-based educational materials and resources is mandatory to achieve the goal of enhancing STEM knowledge and skills in learners with hearing loss. These authors argue that these visual elements are appropriate for the visual abilities possessed by learners with hearing loss. Given that hearing loss creates challenges in mastering spoken and written English or other national languages, it is evident that these visual elements will greatly promote the acquisition of STEM knowledge and skills in learners with hearing loss. There is evidence that in some cases, learners with hearing loss will master materials and resources with the appropriate visual elements in less time than materials or resources lacking those elements (Debevc & Peljhan, 2004). It is clear then that ICT materials and resources that will facilitate the development of STEM knowledge and skills in learners with hearing loss must have these visual elements. Subtitles in English should be mandatory. Other elements such as animations and videos will be equally helpful. The assimilation of educational resources and materials with these visual and other interactive elements by learners with hearing loss will be greatly enhanced (Marschark, et al. 2008).

Clearly, multi-media visual elements will facilitate the acquisition of STEM knowledge and skills by learners with hearing loss in that the materials can be learned repeatedly in the school and at home. Due to legislative mandates, the majority of these learners live at home and attend day schools. Therefore, with ICT materials and resources rich with visual elements, easy and user-friendly interfaces, learners with hearing loss can repeatedly receive support in the school and from family members to learn and practice these materials and resources.

In addition to ensuring that resources and materials to facilitate the development of STEM knowledge and skills in learners with hearing loss have user-friendly visual interfaces such as videos, animation, and subtitles, the use of software specifically developed for this group of learners to facilitate their STEM learning and progress has been documented. For instance, Adamo-Villani and Wilbur (2010) reported two novel approaches to teaching mathematics and science concepts using 3-D animated interactive software for learners with hearing loss.

According to these authors, the content of the software program is based on the academic curriculum. This suggests that this software can address the material learners with hearing loss need to learn at different grade levels. There are animated characters constructed with the latest technology and design that use sign language in this software. This feature is particularly beneficial to learners with hearing loss who use American Sign Language (ASL).

Effective Teaching Strategies and Study Habits on STEM Skills in Learners with Hearing Loss

As stated earlier, parents of children with hearing loss should be supported to understand these children's developmental needs. With this understanding, they could take actions that can lay the foundation for the development of STEM knowledge and skills in their children, for example, by learning how to communicate meaningfully with their children. Ultimately, these children will attend schools, pre-kindergarten (PK) to colleges and universities, where they must acquire deeper STEM knowledge and skills.

Do teachers of learners with hearing loss possess the knowledge and skills to effectively impact STEM knowledge and skills to these learners? This has been and remains an important issue in educating learners with hearing loss. With the enactment of legislation promoting education of learners with hearing loss and other special needs in regular schools in the United States and other Western countries, majority of learners with hearing loss attend regular schools, colleges and universities (Marschark, Sapere, Convertino, & Pelz, 2008). It could be the case that learners with hearing loss will need the support of sign language interpreters or captioning and real-time writers in the regular classroom. Concerns remain that even with these supports, learners with hearing loss in inclusive classrooms may be learning less than their hearing peers (see for example, Marschark, Sapere, Convertino, Seewagen, & Maltzen, 2004).

Experiments conducted by Marschark et al. (2008) indicated that the quality of instruction in terms of accessibility of the material offered to learners with hearing loss in inclusive environments is critically important to how well these students will understand and learn the materials. Given that learners with hearing loss may come to the classroom with lower literacy, metacognition, memory and problem-solving challenges when compared to their normally hearing peers, Marschark, et al. (2008) suggested that to effectively impact the desired knowledge and skills to learners with hearing loss, instructors will have "…exceptional teaching skills beyond their familiarity with the academic abilities and diverse learning styles of deaf students" (p. 558). The question then arises: Do teachers of learners with hearing loss at the various levels of education possess such exceptional teaching skills? Data to address this question are scarce.

Given that the majority of teachers at the various levels of education in inclusive settings may possess very little or no background training and experience in the methods of educating learners with hearing loss, it could be argued that very few teachers in inclusive settings possess such exceptional teaching skills. The implication of this could be that strategies to share information and skills about effective methods for teaching learners with hearing loss will have to be made available to teachers in the regular schools through various in-service training, seminars and workshops on a continuous basis.

Learners ultimately must take responsibility for their learning as they advance through the various levels of education. At the early stages PK – middle grades, a lot of support will be required from the parents or guardians. As learners advance through the various levels of learning, they will need to develop effective study habits in order to learn materials properly, and acquire the necessary knowledge and skills. Sulman and Naz (2012) have argued that effective study habits hold the key to the success of learners with hearing loss in acquiring STEM knowledge and skills. While the literature is replete with tips for effective study habits such as having regular times for study, studying in conducive environments, knowing how to make the best use of resources, making and reviewing notes, etc., the question remains: What is known about the study habits of learners with hearing loss? Data to answer this question are scanty. Learners with hearing loss are different in their degrees of hearing loss, family background, and experiences. Study habits are difficult to examine because as Marchark, Sarchet, Convertino, Borgna, Morrison, and Remet (2012) pointed out it is difficult to determine the veracity of responses to questions on this very personal issue. Nonetheless, evidence indicates that personal factors such as the ability to make effective use of resources such as support services, tutors, interpreters, student peers, and faculty could contribute to the success of students with hearing loss in institutions of higher education (Albertini, Kelly, & Matchett, 2012). Interestingly, areas of weaknesses identified by the deaf students involved in Albertini, et al. (2012) study include their inabilities to prepare for classes, manage their time, concentrate on assignments, identifying important information, preparing for tests and anxiety, motivation, and attitude.

According to Albertini et al. (2012) deaf "students reported high levels of academic difficulty compared to the national norms. They also expressed concern about their study habits, lower levels of verbal confidence, lower motivation to finish college, and less than positive attitudes toward teachers compared to the national norm group of typical entering college students" (p. 99). This implies that deliberate efforts must be made by schools and other educational authorities to explore strategies that will help learners with hearing loss to be aware of their own responsibilities for success in STEM and other areas of academics. Despite availability of abundant technological resources, personal effort is essential to acquire knowledge and develop mastery. It is suggested that adults with hearing loss who are successful in STEM should be identified in the communities.

There are certainly such people. With proper planning, it may be possible for such people to visit educational programs where learners with hearing loss are studying. Meeting, discussing with the students, and responding to their questions could inspire and motivate the students to cultivate the personal habits that will ensure their success in acquiring STEM knowledge and skills.

CONCLUSION

Supporting learners with hearing loss to be successful in STEM must begin immediately following the diagnosis of hearing loss. Given that the vast majority of parents of young children with hearing loss have no previous experience of raising such children, the starting point of the effort to ensure children with hearing loss develop STEM knowledge and skills must be the provision of unbiased and clear information and support to their families. This is necessary to assist the parents and other family members make the necessary adjustments to raising the child with hearing loss. With unbiased and clear information and appropriate supports, the parents could learn about all the resources that could facilitate the raising of their children with hearing loss available to them in the communities. Consequently, they will be aware of their responsibilities in ensuring the success of early intervention programs for the benefit of their children with hearing loss. Research suggests that timely instigation of early intervention programs will ensure that children with hearing loss achieve the same milestones in language development like their hearing peers. Poor literacy skills is one of the greatest challenges to learning STEM materials by learners with hearing loss. Thus the goal of early intervention programs is to ensure these children are successful in language development. In this context 'language', means whatever mode of communication that works for each young person with hearing loss. While a few may, with technology such as cochlear implants, and extensive speech and language therapy services, develop spoken language skills, the majority of children with hearing loss may not develop such language skills regardless of the technologies and services provided. It is very important that professionals in the field provide unbiased and clear information about all the issues in language development of children with hearing loss to their parents. Spoken language is one form of communication. Using manual methods is yet another. The focus must be that the method of communication that is effective with children with hearing loss is consistently used at home and school. This will foster learning and enrich the child's conceptual store. With a consistent medium to communicate and learn, the child with hearing loss will be able to make progress in all aspects of development including acquiring STEM knowledge and skills.

Progress in science and technology is resulting in the development of programs and software rich with visual effects – videos, animations, subtitles, 3-D, etc., that can facilitate the acquiring of STEM knowledge and skills by learners with hearing loss. Nonetheless, given that the majority of children with hearing loss live at homes and attend regular schools due to legislative mandates, it remains a concern whether the teachers in inclusive classrooms are able to utilize these technological resources for the benefit of learners with hearing loss. Many of these teachers may have very little or no background knowledge about educating children with hearing loss and the special ICT resources that could enhance their education. In a very large class in a regular classroom there may be just one or two learners with hearing loss. These students may have a sign language interpreter. The interpreter is a professional trained to facilitate communication between the teacher and students with hearing loss and not a 'teacher'. It remains the task of the teacher to ensure that learners with hearing loss access and comprehend the materials being taught. It is imperative that general education classroom teachers should be made aware of these technological resources that can facilitate the acquisition of STEM knowledge and skills in learners with hearing loss. This can be achieved by regularly conducting in-service training for these teachers by experienced teachers of learners with hearing loss. It is possible that through this activity, teachers in inclusive programs will become very aware of the unique needs of students with hearing loss in their classrooms and always plan to address such needs. Otherwise, there is real danger that the needs of learners with hearing loss in inclusive classrooms will not be met. Consequently, acquiring the knowledge and skills in STEM and other academic areas will be problematic.

Finally, as learners with hearing loss advance through the various levels of education, it may be necessary for them to be provided mentors who will be working with them and reinforcing the development of effective study skills. Intrinsic motivation is necessary for academic success. Thus some learners with hearing loss may benefit if they have mentors who check-in with them, discuss their challenges and progress. Especially beneficial will be mentors who are successful in STEM and who have hearing loss themselves. These will make wonderful role-models as well as inspire and encourage young learners with hearing loss. It may be the case that education authorities will set up a program focused at identifying such successful role-models with hearing loss. With proper planning, many of such people may be glad to visit schools, meet and discuss with learners with hearing loss on a voluntary basis.

Taken together, learners with hearing loss can, with the appropriate supports starting immediately after the diagnosis of hearing loss, acquire STEM knowledge and skills. History is replete with successful people in these fields who had hearing loss. Given continuing progress in science and technology and development of innovative, user-friendly, multi-media educational programs and materials almost on a daily basis, learners with hearing loss can be very successful in acquiring STEM knowledge and skills and be able to contribute to the development of society.

References

Adamo-Villani, N., & Wilbur, R. (2010). Software for math and science education for the deaf. *Disability and Rehabilitation: Assistive Technology, 5*(2), 115-124.

Albertini, J. A., Kelly, R. R., & Matchett, M. K. (2012). Personal factors that influence deaf college students' academic success. *Journal of Deaf Studies and Deaf Education, 17*(1), 85-101.

Blatto-Vallee, G., Kelly, R. R., Porter, J., & Fonzi, J. (2007).Visual-spatial representation in Mathematical problem solving by deaf and hearing students. *Journal of Deaf Studies and Deaf Education, 12*(4), 432-448.

Borum, Y. (2012). Perceptions of communication choice and usage among African American hearing parents: Afrocentric cultural implications for African-American deaf and hard of hearing children. *American Annals of the Deaf, 157*(1), 7-15.

Calderon, R. (2000). Parents' involvement in deaf children's education programs as predictors of child language, early reading, and social-emotion development. *Journal of Deaf Studies and Deaf Education 5*, 140–155.

Coppens, K. M., Tellings, A., Schreuder, R., & Verhoeven, L. (2013). Developing structural model of reading: The role of hearing status in reading developing over time. *Journal of Deaf Studies and Deaf Education, 18*(2), 1-24.

Davila, R. R. (2004). *Reviewing the past, assessing the present, and projecting the future*. In D. Powers, & G. Leigh, (Eds.). Educating deaf children: Global perspective, pp 3–12. Washington, DC: Gallaudet University Press.

Deaf Scientists Corner. (2013). Deaf scientists' corner: A website devoted to the biographies of famous deaf scientists. Retrieved from: http://www.twu.edu/dsc/index.htm

Debevc, M., & Peljhan, Z. (2004).The role of video technology in on-line lectures for the deaf. *Disability & Rehabilitation 26*(17), 1048-1059.

Eleweke, C. J., Bays, D., Gilbert, S., & Austin, E. (2008). Information about support services for families of young children with hearing loss: A review of some useful outcomes and challenges. *Deafness and Education International, 10*(4), 190 -212.

Fulcher, A., Purcell, A. A., Baker E., & Munro, N. (2012). Listen up: Children with early identified hearing loss achieve age-appropriate speech/language outcomes by 3 years-of-age. *International Journal of Pediatric Otorhinolaryngology 76*(12), 1785-1794.

Gomez, G. M., Piehota, L. D., & Dischner, R. R. B. (2012). Providing a program to meet the needs of toddlers with hearing loss and deafness. *Perspectives on Hearing & Hearing Disorders in Childhood. 22*(1), 4-10.

Holzinger, D., Fellinger, J., & Beitel, C. (2011). Early onset of family-centered intervention predicts language outcomes for children with hearing loss. *International Journal of Pediatric Otorhinolaryngology, 75*(2), 256-260.

Hyde, M. B., & Power, D. J. (2003). Characteristics of deaf and hard of hearing students in Australian regular schools: Hearing level comparisons. *Deafness & Education International, 5*, 133-143.

Jackson, C. W. (2011). Family supports and resources for parents of children who are deaf or hard of hearing. *American Annals of the Deaf, 156*(4), 343-362.

Jamison, J. R., Zaidman-Zait, A., & Poon, B. (2011). Family support as perceived by parents of preadolescents and adolescents who are deaf or hard of hearing. *Deafness & Education International, 13*(3), 110-130.

Kasai, N., Fukushima, K., Omori, K., Sugaya, A., & Ojima, T. (2012). Effects of early identification and intervention on language development in Japanese children with prelingual severe to profound hearing impairment. *The Annals of Otology, Rhinology & Laryngology. Supplement, 202*, 16-20.

Marschark, M. (2007). Raising and educating a deaf child: A comprehensive guide to the choices, controversies, and decisions faced by parents and educators. New York: Oxford University Press.

Marschark, M., Lang, H. G., & Albertini, J. A. (2002). Educating deaf students: From research to practice. New York: Oxford University Press.

Marchark, M., Sarchet, T., Convertino, C. M., Borgna, G., Morrison, C., & Remet, S. (2012). Print exposure, reading habits, reading achievement among deaf and hearing college students. *Journal of Deaf Studies and Deaf Education, 17*(1), 61-74.

Marschark, M., Sapere, P., Convertino, C., & Pelz, J. (2008). Learning via direct and mediated instruction by deaf students. *Journal of Deaf Studies and Deaf Education, 13*(4), 546-561.

Marschark, M., Sapere, P., Convertino, C., Seewagen, R., & Maltzen, H. (2004). Comprehension of sign language interpreting: Deciphering a complex task situation. *Sign Language Studies, 4*, 345-368.

McKay, S., Gravel, J. S., & Tharpe, A. M. (2008). Amplification considerations for children with minimal or mild bilateral hearing loss and unilateral hearing loss. *Trends in Amplification, 12*(1), 43-54.

Mitchell, R. E., & Karchmer, M. A. (2004). Chasing the mythical ten percent: Parental hearing status of deaf and hard of hearing students in the United States. *Sign Language Studies, 4*(2), 138-163.

Moores, D. F. (2010). Epistemologies, deafness, learning and teaching. *American Annals of the Deaf, 154*(5), 447-455.

Mukuria, G. M., & Eleweke, C. J. (2010). Educating children with deafness and hearing impairments. In: P. Penelope, B. Eva, & M. Barry, (Eds.), *International encyclopedia of education, 2*, 628-633. Oxford, England: Elsevier.

Scheetz, N.A. (2012). *Deaf education in the 21st Century: Topics and trends*. Boston, MA: Pearson.

Spencer, E. P., & Marschark, M. (2010). Evidence-based practices in educating deaf and hard of hearing students (Professional perspectives on deafness: Evidence and applications). New York: Oxford.

Sulman, N., & Naz, S. (2012). Relationship between study habits of deaf students and their performance in general science. *Interdisciplinary Journal of Contemporary Research in Business, 3*(12), 489-495.

Turnbull, A. P., & Turnbull, H. R. (2006). *Families and exceptionalities, 5th edition*. Upper Saddle River, NJ: Pearson/Merrill/Prentice Hall.

Turnbull, A. P., Turnbull, A., Shank, M., & Smith, S. J. (2004). *Exceptional lives: Special education in today's schools, 4th edition*. Upper Saddle River, NJ: Pearson/ Merrill/Prentice-Hall.

Vosganoff, D., Paatsch, L. E., & Toe, D. M. (2011). The mathematical and science skills of students who are deaf or hard of hearing educated in inclusive settings. *Deafness & Education International, 13*(2), 70-88.

Yoshinaga-Itano, C. (2004). Levels of evidence: universal newborn hearing screening (UNHS) and early hearing detection and intervention systems (EHDI). *Journal of Communication Disorders 37*, 451–465.

In: S.T.E.M. Education
Editor: Satasha L. Green

ISBN: 978-1-63117-806-1

Chapter 10

WORKING WITH LEARNERS WITH VISUAL IMPAIRMENTS IN STEM

Kimberly M. Edwards and Satasha L. Green*
Chicago State University, IL, US

ABSTRACT

When considering the various representations of disabilities within a classroom, teachers often struggle to provide adequate instruction for students with visual impairments. Of the primary senses, visual acuity serves as a significant instrument in K-12 learning processes. Students with low vision, blindness, or other visual impairments must find ways to compensate for their respective limitation. Consequentially, within the classroom, teachers also have an imperative responsibility of assisting students in the learning process by providing prescribed accommodations and giving the necessary guidance and support for every student to achieve success within the learning process.

Many teachers have noted the challenges of understanding and implementing prescribed accommodations within the classroom. Moreover, these tasks may seem particularly ominous within STEM-related courses, especially within mathematics and the sciences, as these fields contain many courses that have been traditionally considered "visually dependent" subject matter. As of recent, the numbers of students with visual impairments entering into STEM-related courses in higher education and STEM fields upon graduation has been slowly but steadily increasing. This increase can be credited to the attitudes, support and preparation of K-12 students in math and science. As mathematics and science at the K-12 level continue to serve as the "gatekeepers" to entry into STEM programs in higher education, teachers in K-12 classrooms are charged with the task of preparing students in these subjects, not only through traditional pedagogy, but through a clear understanding of the needs of all students.

* sgreen34@csu.edu.

INTRODUCTION

Within today's classroom, learning takes place in a variety of ways. Within these modes, students bring a wide range of learning styles and rely heavily on their senses to process course material. Of our five primary senses, vision and visual acuity have an imperative role in the learning process within "traditional" learning environments. For example, many times the ability to read textbooks or see written information on whiteboards is assumed to be necessary for successful learning in the classroom. Therefore, the presence of visual impairments must be acknowledged and effectively addressed within today's classroom.

Within children, vision loss can be caused by damage or injury to the eye, incorrect shaping of the eye, or issues within the brain (Center for Disease Control, 2013). The most common eye disorders include refractive errors including myopia (near-sightedness), hyperopia (far-sightedness) and astigmatism (Center for Disease Control, 2013). Most often, these disorders can be corrected through surgery or the use of corrective lenses or contact lenses. Yet, the range of visual impairments sometimes extends further than mild vision loss or changes. There are a number of neurological and congenital impairments of the eye, which also results in sensitivity to light, color blindness, structural damage to the eye, depth perception and acuity of vision (APH, 2012). When considering these limitations, there are numerous definitions which address a range of visual impairments, ranging from mild to severe limitations and originating from both educational and legal fields.

As there are various causes of visual impairments that occur within childhood, the degree of impairment becomes a major factor in defining the impairment itself. The degree of impairment is heavily weighted within defining the eye condition, a visual process that is affected, and the ability to correct the limitation (NDCCD, 2012). Within education, one of the primary methods of assessing and defining visual ability is based on the ability to read. For example, if an individual with partial vision can learn to read through modifications such as using large print or a magnifier, this individual is often classified as being legally blind with low vision (APH, 2012). A child who is able to read through Braille or by using auditory methods and has no vision is referred to as being totally blind (APH, 2012). Within the medical field, there are various ways to define the scope of visual ability, as well. For example, the World Health Organization (WHO) cites four levels of visual functioning: (1) normal vision, (2) moderate visual impairment, (3) severe visual impairment, and (4) blindness (2013).

According to the U.S. Department of Education (2014), children with visual impairments comprise one of the smallest categories within the scope of disabilities, representing approximately .04% of school-aged children. Although few in numbers, the severity found within aspects of this disability can definitely challenge both educators and students within teaching and learning processes. The scope of this definition extends well beyond mild visual impairments that are corrected with glasses or contact lenses. In the United States, there are approximately 60,000 children who are classified as legally blind (APH, 2012), with legal blindness being defined as having a "central visual acuity of 20/20 or less in the better eye with the best possible correction, or a visual field of 20 degrees or less" (AFB, 2013). This can include a child that has very limited vision, with some measurable level of visual acuity, or a child that has total blindness; resulting in approximately 30,000 children between the ages of 3 and 21 receiving services within schools under the mandates of the Individuals with Disabilities Education Act (34 C.F.R. § 300.346),

Within education, one of the most widely known definitions is the legal definition contained within IDEA. As mandated by IDEA, states are required to annually report the number of children with visual impairments who receive special services (IDC, 2012). As such, visual impairment is defined as "an impairment in vision that, even with correction, adversely affects a child's educational performance. The term includes both partial sight and blindness" (U.S. Department of Education, 2014). It is through this definition that children receive services and support for a myriad of visual impairments in schools across the country. Parents of children with visual disabilities can have the child evaluated, without cost, to determine the need of special services under IDEA. Special services can include early intervention if the limitation is detected before the child's third birthday or special education and other related services if a child is between the ages of three and twenty-one (NDCCD, 2012).

IMPORTANCE OF STEM INSTRUCTION FOR LEARNERS WITH VISUAL IMPAIRMENTS

Research strongly suggests that success in STEM careers is positively correlated with ample preparation in mathematics and science within K-12 education (NSF, 2012). Science and mathematics within K-12 education often serve as a conduit into higher education and success in STEM majors. Unfortunately, Americans with disabilities have historically been excluded and underrepresented within STEM education (Fitchen, Baile, & Asuncion, 2003) and must often overcome barriers from a wide variety of sources. Most prevalent are those obstacles found in the classroom structure for both the students with a visual impairment seeking STEM-related courses and teachers and school personnel. The complexities of providing STEM instruction is often noted within all grade levels. As students enter higher grades, providing accommodations within STEM instruction becomes more complicated as courses and course content become more difficult (Fitchen et al., 2003). At these levels, STEM education is often combined with lab activities and other activities outside the classroom, which increase the complexity found within instruction of students with visual impairments. In many cases, "hands-on" often equates to "eyes on" within STEM instruction, as in large proportion of laboratory-based, technology-based, and equation-based instruction depends heavily on visual acuity (Fitchen et al., 2003). Therefore, the misconception that the students cannot successfully navigate within STEM-focused environments serves as the foundation for biases that unfairly support limitations of the students in these environments.

Students with visual limitations often experience a lack of encouragement in their pursuit of completing STEM-related courses from educational and familial systems that lack knowledge on how to effectively support these learners. According to the National Science Foundation (2002), many teachers and families hold the misconception that students with disabilities cannot be successful in STEM-related fields. In many cases, this fallacy often develops from a perceived need to protect the student from some anticipated failure or harm. Consequentially, many students with visual impairments are dissuaded from entering advanced math and science courses that often lead to acceptance in STEM programs in higher education later in life. Students with visual disabilities often encounter low expectations from teachers and family, limited exposure to STEM related courses, lack of role models within STEM fields and an overall lack of support within their STEM education (Dunn et al., 2012). When they do enter into these classes, often times they are not fully integrated into the rigors of learning activities such as laboratory-based assignments (Fitchen et al., 2003). Often students in mathematics and science courses often remain segregated within the general classroom and are found within classrooms that do not prepare for the rigors of STEM education (Fitchen et al., 2003). Support systems for students with visual impairments can become a hindrance if these supports underestimate students' abilities. It is recommended that teachers be well aware of explicit and implicit bias towards students with disabilities, particularly in the early stages of student education (Dunn et al., 2012). It is during this time that interests in STEM education can be either nurtured or diminished, largely impart through educator attitude.

Another challenge to educators is the inability to effectively identify needed accommodations within the classroom. Rule, Stefanich, Haselhuhn, and Peiffer (2009) noted that many teachers are not prepared to provide meaningful instruction to students with disabilities in math and science courses. This deficit includes a limited understanding of how to modify instruction, differentiating instructional materials, and lack of overall understanding in identifying and providing appropriate accommodations within the classroom. Children with visual impairments rely on the ability of the teacher to provide differentiated instruction, identify and supply needed accommodations, and provide the appropriate level of support. Unfortunately, teachers often find themselves unprepared and ill-equipped to address the needs of students with visual impairments within STEM instruction (Fitchen et al., 2003). Further, they may not have access to the resources needed for accessible teaching.

The level of modification and accommodation students should receive is detailed under IDEA. Parents and program professionals must work closely to develop an Individualized Education Program (IEP), which indicates the explicit need for such unified support and planning. Yet, teachers often find themselves struggling to understand the implementation of the modification in the classroom or find it cumbersome to differentiate instruction in the manner required. Alston and Hampton (2000) further noted that students with disabilities often receive unfair treatment within the classroom by both teachers and fellow classmates due to the often negative views of the accommodation itself, with many feeling that accommodations such as extra time to take a test is an unfair advantage or providing instruction in more than one manner is a nuisance.

Pedagogical Approaches to STEM for this Population

One common misconception is students with one or more disabilities often struggle in the integrated classroom. Contrarily, research asserts that students with disabilities within integrated classrooms often outperform their counterparts within segregated classrooms (Ferguson& Asch, 1989; Baker, Wang, & Walberg, 1994). It must be recognized that students with visual impairments have the same cognitive abilities as their sighted counterparts. Within the learning process, the understanding can be the same with students with visual impairments and student who are fully sighted. Yet, it is the method of gaining understanding that can differ greatly within individuals. Even within a classroom of individuals with visual impairments, individuality within learning processes will produce a wide range of variances within the classroom. Kumar, Ramasamy, and Stefanich (2001) noted, while visual impairments are likely to prohibit individuals from variable experiences, conquering obstacles to experiencing activities that are unknown is critical in motivating the intellectual growth of students with visual impairments.

Students with visual impairments have most likely learned to utilize other sensory perceptions in the development of their specific learning style. Important tasks for teachers attempting to provide instruction to students with visual impairments are gaining an understanding of the students' limitations, gauging specific learning styles utilized by students, and understanding prescribed accommodations. In order to effectively understand necessary accommodations, teachers must be familiar with the requirements within the student's IEP, as well as gain a clear understanding of the accommodations needed through the student. The U.S. Office of Civil Rights (2011) asserted the importance of students' with disabilities aptitude to identify and understand the limitations within their disability and, moreover, competence in verbalizing their specific needs for accommodation. Although this may be a difficult task for younger children, parents in conjunction with educators should have constant and consistent paths of communication regarding these needs. They must also monitor the effectiveness of accommodations through frequent evaluation at various levels.

Within general instruction, teachers should be prepared to modify instruction when a child with a visual impairment is present in the classroom. This should not be mistaken with eliminating parts of the curriculum or standards within instruction. Instead, teachers have the responsibility of understanding the individual needs of a student with a visual impairment within the classroom and find creative and instructive methods of modification for presentation of materials.

K-12 Teacher Strategies and Resources for Working with this Population in STEM

Differentiated instruction recognizes the variances within abilities and learning styles within student populations. Differentiated instruction involves providing activities and lessons that are understood by students representing varying levels of ability and skill sets. Within the context of differentiated instruction, teachers are responsible for assessing learning style variances and modifying teaching strategies accordingly. Forsten, Grant and Hollas (n.d.) have developed a comprehensive list of "building blocks" needed for successful differentiated instruction. These recommendations include:

Knowing the Learner. Teachers need to understand who their students are and their unique qualities including learning styles, personality, interests, strengths, limitations, levels of ability and family status.

Quality Teacher. A quality teacher believes all students can learn, is able to support student learning, is committed to ensuring that all students learn, and is able to identify both what works and what does not.

Quality Curriculum. A quality curriculum is of interest to students is thoughtful and provoking, relevant to students, focused on quality and sequenced to optimize learning.

Classroom Learning Environment. A quality classroom environment provides seating and accommodations based on student needs, uses flexible grouping, provides appropriate resources, and has a balanced population.

Flexible Teaching and Learning Time Resources. The best use of time within instruction and learning includes team teaching, tutoring and remediation and additional resources for needed instructional support.

Evaluation Assessment and Grading. Evaluation should include the consideration of numerous ways to assess learning. Oral reports, written reports, drawings, and performances are all helpful.

Knowing and understanding these principles is useful within any classroom and, specifically, serve as useful tenets for teachers as they prepare to instruct students with visual impairments. Within STEM instruction, there are some modifications that can be universal across STEM-related courses, yet, unique to individuals with visual impairments within these courses. STEM courses usually consist of information conveyed within visual simulations of charts, diagrams, drawings, graphs, and 3-D representations (Jones, Minogue, Oppewal, Cook, & Broadwell, 2006). It is important to understand that students with low vision often find it difficult to view this information within PowerPoint and whiteboard presentations. Therefore, students with visual impairment most often rely heavily on verbal description as a means of gaining conceptualized understandings of concepts within the classroom. Teachers need to remain conscious of this reliance and use paraphrasing and reflective activities to gauge the level of understanding of the material being presented. Providing clear verbal description includes the elimination of vague concepts such as "this" or "those" (Dion, Hoffman & Matter, 2000). Through clear explanation, materials and concepts should be explained using detailed terminology and description, thereby, allowing students to gain a clear visual interpretation and shared meaning of that which is being discussed. Although verbal description is helpful, teachers should also provide three-dimensional objects to support understanding, whenever possible.

Children with limited vision or blindness should be given the resources to effectively master the core curriculum at the same rate as sighted peers, although mastery may not be achieved in the same manner. Differentiated instruction allows for the exploration of proven and creative ways to effectively construct learning within a variety of ways to many different individuals. Teachers are often able to effectively apply the theory of differentiated instruction to students with varying levels of abilities and limitations, including a learner with a visual impairment who may be limited in vision or have no vision at all. Differentiated instruction is most beneficial in this case, as it creates the unique and mutually beneficial opportunity for partnering with parents, the IEP team and the student to create an instructional plan that best develops successful opportunities for learning and student growth. It is within this process of building and developing that support systems are effectively constructed to support the student with a visual impairment in the learning process.

Modifications for students with visual impairments vary depending on the type and severity of the impairment combined with the individual needs of each student. Modifications may include using Braille to read and write, individualized instruction on how to use specialized computer equipment and other devices designed to assist individuals with visual impairment; and, if some vision is present, using the vision they have in an effective manner. Materials within the course are very important for students with visual impairments.

Whenever possible, handouts and notes should be distributed beforehand, including any required accommodation. Teachers should also ensure that any technology or other resources needed are available well before the class begins. It is common for students to have special textbooks (i.e. written in Braille or in large print) and note takers within the classroom. Further, students may have access to technology such as magnifiers, voice dictation software, transcription hardware, and text readers. In Table 1 Wiazowski (2009) provided the following resources as possible considerations for assistive technology for students with visual impairments.

Table 1. Assistive Technology for Students with Visual Impairments

• Text-to-Braille translation software or other software to translate print to Braille.
• An embosser (Braille printer).
• Scanner with Optical Character Recognition (OCR) software.
• Image simplifying software
• Image embossing devices
• Color copier with a magnification function
• Text-to-talk software
• Voice recording/dictation software

There are also 'smart phone' applications available with many of these functions. Teachers should gain some level of working knowledge of these accommodations, especially those related to technology. Teachers often serve as the first line of support in the classroom when these accommodations do not function as required. Having some knowledge of these allows teachers to generate quick solutions to potential issues often without a lengthy disruption within classroom instruction. Based upon Kumar et al. (2001) suggestions for teachers, below in Table 2 are some possible modifications within a classroom.

Table 2. Classroom Modifications

• Provide course materials into Braille and adaptive electronic media.
• Distribute course materials and notes to students before class.
• Provide materials in advance, whenever possible.
• Allow students to audiotape classroom lectures.
• Refrain from using vague terminology and be specific when giving directions.
• Provide students with low vision large print copies of materials.
• Provide adaptive software and hardware, as needed.
• Provide a wide range of tactile learning experiences, whenever possible.
• Provide ample space should a student have a guide dog and ensure that other students understand the role of the animal.
• Pair the visually impaired student with a classmate to read directions and procedures, and guide him/her through activities. Provide assistive technologies whenever possible.

One factor that is often overlooked when considering the instruction of students with visual impairments is creating a physical environment that is conducive for navigation. Students with visual impairments rely heavily on the classroom teacher to create a learning environment that provides a sense of safety, security and familiarity. Teachers should take the time to provide both a verbal and tactile tour of the classroom for students entering the learning environment for the first time. Once students become familiar within their environment, changes should be limited, and if necessary, clearly explained to students. In doing such, students with visual impairments are able to gain and maintain a sense of navigational independence within the learning environment.

Differentiating STEM Instruction for this Population of Learners

Within K-12 education, mathematics and the sciences are the most common STEM-related courses found within educational settings. Unfortunately, teachers have had few models for teaching students with disabilities in the sciences and mathematics and there is very little research on interventions in these areas (Browder, Trela, Courtade, Jimenez, Knight, & Flowers, 2012). Engineering and technology education are usually offered within higher education instruction. There are some technology-focused courses within K-12 education; yet, the concept of providing technology-focused courses to visually impaired students is often intertwined with the learning and usage of assistive technology in the classroom.

Technology-focused courses within STEM go beyond instruction on assistive technology. Assistive technology is equipment or systems used to improve the capabilities of individuals with disabilities (Moon, Todd, Morton, & Ivy, 2012). Although understanding assistive technology is important to students with visual limitations, teaching a student with a visual disability to use assistive technology does not equate to teaching technology focused courses within them. Technology courses within STEM instruction should have well-defined curricula. Whereas assistive technology skills can effectively benefit any facet of related education, technology courses within STEM most often focus on a computer or technology-based curriculum within instruction. Therefore, much of the differentiation within instruction is found within mathematics and the sciences within K-12 education.

Differentiated Science Instruction

The sciences serve as the foundation of successful entry into STEM careers. However, students with disabilities, including visual impairments, are not readily accommodated within science instruction (Kumar et al., 2001). Specifically, with instruction for students with visual impairments, there is often some apprehension from science teachers regarding the laboratory experiences that are often required within many science-focused courses. This is often due to the many perceived safety hazards found in laboratories. However, Fraser and Maguvhe (2008) provided an effective and insightful reminder to educators that, in order for STEM education to be effective for all students, laboratory sciences should be multi-sensory in nature. Specifically, this entails the use of other senses when sight is limited or absent. When considering the many safety hazards present in the laboratory for all students, special attention should be the priority for all, but especially those with visual impairments. Yet, these hazards should in no way serve as an excuse to limit learning experiences for a student with a visual impairment.

Chemical identification is a great safety concern for many students. Within a laboratory housing a student with a visual impairment, chemicals should be organized in the same place, on the same shelf and labeled in an accessible way which may include using large print or Braille for labeling (Wedler, Boyes, Davis, Flynn, Franz, Hamann, & Wang, 2013). Special sensors can be purchased to prevent overfilling containers in the lab and electrode meters are available to obtain accurate measurement within container (Moon et al., 2012). Further, equipment instruction and key documents within the laboratory should be available in large print or Braille formats.

Heat sources within the lab are of concern for all students, as there is the potential for flammable material to come in contact with the heat source. Hot plates have been identified as viable alternatives to Bunsen burners. Further, for low vision students, tripod stabilization helps to secure the Bunsen burner in place to prevent accidentally tipping over the heat source (Moon et al., 2012). Auditory senses become important to students with visual impairments in the lab as well. It can allow for the identification of heat sources and serve as a way to notify the student of changes in their own heat source, including the recognition of a lit and extinguished heating source. Additional resources for science instructional practices are available via the use of haptic aids and computers, allowing visual learners to become more actively involved. The haptic instructional program has been used in a series of studies that allow students to feel particle movement in a closed system (Jones, Childers, Emig, Chevrier, Tan, Stevens, & List, 2014).

The concept of particle movement is crucial for individuals to understand various interdisciplinary science concepts, such as heat; the formation of viral capsids, proteins, and structures; and processes such as osmosis (Jones et al., 2014). Through technology, companies have developed equipment such as talking thermometers, calculators, and microscopes that allow for greater independence in the laboratory (Rule, Stephanic, Boody, & Peiffer, 2011). Through practice, students can learn to perform within the laboratory in an effective and efficient manner.

Within the laboratory, a sighted partner also can become a resource to a student with a visual impairment. Although widely proven that students with visual impairments can successfully navigate within the laboratory setting successfully without assistance, having a sighted partner can provide additional details for understanding that a student with a visual impairment may not be able to capture. For example, a student who is blind can set up a laboratory experiment and their partner can enter data about the finding.

Within the science classroom, students often do not have the ability to gain understanding of concepts fully through verbal or two-dimensional descriptors. Companies have responded to this by creating various three-dimensional models that can be used in science courses. For example, several companies create three-dimensional models related to anatomy or biology (Garrido-Escudero, 2013) which are helpful in using tactile senses to create understanding of concepts, such as the parts of the cell or the anatomical construction of a hand. Teachers should also ensure that the proper technology is in place within the classroom for students with visual impairments.

This includes large-screen monitors for classroom presentations. Also, if using visual aids, such as movies or videos, students with visual impairments should be provided resources to access these aids for further viewing outside the classroom.

Differentiated Mathematics Instruction

Teaching mathematics to students with disabilities often creates a challenge within instruction of students with visual impairments due to the visible nature of math instruction. Although challenging, with modification, instruction to students with visual impairments can prove to be highly rewarding. It becomes important to understand that students with visual impairments should not be excused from math solely because of a visual impairment. Teachers must be willing to teach mathematics to all students within the class, using the best possible resources to do so.

Younger children benefit greatly from the use of tactile within the classroom. Items such as blocks, Legos, abaci, weights and other three-dimensional objects assist in understanding numbers, quantity, volume, size, and weight (Fichten et al., 2003). Also, explaining concepts in song serves as a creative way to present concepts in a manner through which they are remembered. Teachers must also recognize the potential for hidden assumptions. For example, students may not possess knowledge of things such as complex shapes or of spatial concepts. Before students with visual impairments can begin to understand mathematics, conceptually, on paper, they should have an opportunity whenever possible to gain a clear understanding of mathematics from a tactile perspective as explained by the math teacher. Thus, math teachers play a central role in providing clear and appropriate vocabulary for understanding within the classroom.

Technology plays a large part in more complex mathematics classes offered to children. In the preparation of class documents for instruction, teachers should ensure the use of larger fonts on worksheets and handouts for student with low vision. For students with complete blindness, audio and visual presentations can assist in explaining complex concepts and equations. If information cannot be easily conveyed through verbal description, raised-line drawings become options in subjects such as geometry and others where graphs are prevalent. Technology has produced a wide variety of equipment, such as talking calculators that allow for a greater understanding and navigation of mathematical equations (Fichten et al., 2003). Today, much technology is often found within standard packages on many personal computers. Visual magnification, voice dictation and voice readers, and screen magnification are readily available applications for mathematics instruction.

Especially helpful within higher-level coursework, students with visual impairments often used the Nemeth Code for Braille Mathematics and Science Notation (Nemeth code). Nemeth code uses standard Braille to create an understanding of mathematics, including mathematical equations within the sciences (Nemeth, 2001). Yet, Rosenblum and Amato (2004) noted the problems using instruction with Nemeth code, as teachers often do not have the skills necessary to prepare the materials or give adequate instruction. Further, there are often discrepancies within materials prepared from textbooks into Braille. Despite these issues, Nemeth code has been found to be helpful within a variety of mathematics courses including arithmetic, long division, algebra, geometry, trigonometry, and calculus (Rosenblum & Amato, 2004).

CONCLUSION

In 2013, President Barack Obama asserted the declaration, "If we want America to lead in the 21st century, nothing is more important than giving everyone the best education possible — from the day they start preschool to the day they start their career." Within the spirit of this, educators must be mindful that "everyone" includes students with disabilities and specifically those students with visual impairments. As jobs in STEM-related fields continue to grow in abundance, supporting students with visual impairments sets the foundation for success in the pursuit of these careers within the long-range plans of these students. In order to effectively do so, educators should be mindful of their responsibilities within the classroom in providing effective support. Taking deliberate actions to understand student limitations and needs within accommodations creates a foundation that supports the successful implementation of STEM-related courses into course planning of students with visual impairments.

Given the many barriers that already face students with visual disabilities, educators have the unique responsibility to provide positive support and guidance to students interested in pursuing STEM careers and wish to enroll in STEM-related courses. Teachers should continuously assess the effectiveness of accommodations and provide a level of instruction that supports the visually impaired student and allows for the students and others to realize that a visual impairment does not equate to total limitations within achieving success in STEM fields.

To effectively teach a student with a visual impairment, teachers must be willing to differentiate instruction and adapt to the individual learning styles of each student. Although challenging, instructing a student with a visual impairment can also be rewarding for both the teacher and the student. Full participation can be achieved through the implementation of individualized teaching methods and assistance. Not only do teachers have the responsibility of assisting students with visual impairments and gaining an understanding of the STEM-focused classroom, there is also an overarching responsibility for teachers to support future successes in the STEM fields as students with visual impairments move forward, thus, becoming a catalyst in providing unlimited possibilities found within STEM education.

REFERENCES

Alston, R. J., & Hampton, J. L. (2000). Science and engineering as viable career choices for students with disabilities: A survey of parents and teachers. *Rehabilitation Counseling Bulletin, 43*(3), 158-164.

American Foundation for the Blind. (2014). *School experience for children and youth with vision loss*. Retrieved from http://www.afb.org/info/blindness-statistics/children- and-youth/school-experience/235.

The American Printing House for the Blind. (2012). *Annual report.* Retrieved from http://www.aph.org/annual-reports/2011/.

Baker, E., Wang, M., & Walberg, H. (1994). The effects of inclusion on learning. *Educational Leadership, 52*(4), 33-35.

Browder, D. M., Trela, K., Courtade, G. R., Jimenez, B. A., Knight, V., & Flowers, C. (2012). Teaching mathematics and science standards to students with moderate and severe developmental disabilities. *The Journal of Special Education*, *46*(1), 26-35.

Centers for Disease Control and Prevention. (2013). *Vision health initiative: Common eye disorders*. Retrieved from http://www.cdc.gov/visionhelath/basic_information /eye_ disorders.htm

Dion, M., Hoffmann, K., & Matter, A. (2000). *Teacher's manual for adapting science experiments for blind and visually impaired students*. Worcester Polytechnic Institute.

Dunn, C., Rabren, K. S., Taylor, S. L., & Dotson, C. K. (2012). Assisting students with high-Incidence disabilities to pursue careers in science, technology, engineering, and mathematics. *Intervention in School and Clinic*, *48*(1), 47-54.

Ferguson, P., & Asch, A. (1989). Lessons from life: personal and parental perspectives on school, childhood, and disability. In Biklen, D., Ford, A., & Ferguson, D. (eds.), *Disability and Society*. Chicago: National Society for the Study of Education, pp. 108-40.

Fichten, C., Barile, M., & Asuncion, J. (2003). Computer technologies and postsecondary students with disabilities: Implications of recent research for rehabilitation psychologists. *Rehabilitation Psychology 48*(3): 207-214.

Forsten, C., Grant, J., & Hollas, B. (n.d.). The 7 building blocks of differentiated instruction. DI.

framework in progress. Retrieved from http://www.sde.com/Downloads/ TeacherResources/ di/7_building_blocks.pdf

Fraser, W. J., & Maguvhe, M. O. (2008). Teaching life sciences to blind and visually impaired learners. *Journal of Biological Education*, *42*(2), 84-89.

Garrido-Escudero, A. (2013). Using a hands-on method to help students learn inorganic chemistry nomenclature via assembly of two-dimensional shapes. *Journal of Chemical Education*, *90*(9), 1196-1199.

Individuals with Disabilities Education Act Amendments of 1997, Pub. L. No. 105-17, 111 Stat. 37 (1997).

IDEA Data Center. (2012). *Public Data and Resources*. Retrieved from http://www. ideadata.org/tools-and-products.

Jones, M. G., Childers, G., Emig, B., Chevrier, J., Tan, H., Stevens, V., & List, J. (2014). The efficacy of haptic simulations to teach students with visual impairments about temperature and pressure. *Journal of Visual Impairment & Blindness*, *108*(1), 55-61.

Jones, M. G., Minogue, J., Oppewal, T., Cook, M. P., & Broadwell, B. (2006). Visualizing without vision at the microscale: Students with visual impairments explore cells with touch. *Journal of Science Education and Technology 15*(5/6), 345-351.

Kumar, D., Ramasamy, R., & Stefanich, G. P. (2001). Science for students with visual impairments: Teaching suggestions and policy implications for secondary educators. *Electronic Journal of Science Education*, *5*(3). Retrieved from http://ejse. southwestern.edu/article/view/7658/5425

Moon, N. W., Todd, R. L., Morton, D. L., & Ivey, E. (2012). Accommodating students with disabilities in science, technology, engineering, and mathematics (STEM): Findings from research and practice for middle grades through university education. Atlanta: Center for Assistive Technology and Environmental Access, College of Architecture, Georgia Institute of Technology. Retrieved from http://www.catea.gatech.edu/scitrain/ accommodating.pdf.

National Dissemination Center for Children with Disabilities. (2012). NICHCY Disability Fact Sheet 13. Retrieved from http://nichcy.org/

National Science Foundation. (2002). *NSF's program for persons with disabilities: A decade of innovation and progress.* Retrieved from http://www.nsf.gov/ pubs/2002/ nsf02094/nsf02094.pdf

Nemeth, A. (2001), The Nemeth Code. In Dixon, J. M. (ed.), *Braille: Into the Next Millennium.* National Library Service for the Blind and Physically Handicapped, Washington, D.C, p.120-127.

Rosenblum, L. P., & Amato, S. (2004). Preparation in and use of the Nemeth Braille Code for mathematics by teachers of students with visual impairments. *Journal of Visual Impairment & Blindness, 98*(8).

Rule, A. C., Stefanich, G. P., Haselhuhn, C. W., & Peiffer, B. (2009). *A working conference on students with disabilities in STEM coursework and careers.* Proceedings from the 2009 Working Conference on Students with Disabilities in STEM Coursework and Careers, Cedar Falls, IA. (ERIC Document Reproduction Service No. ED505568)

Rule, A. C., Stefanich, G. P., Boody, R. M., & Peiffer, B. (2011). Impact of adaptive materials on teachers and their students with visual impairments in secondary science and mathematics classes. *International Journal of Science Education, 33*(6), 865-887.

U.S. Office of Civil Rights. (2011). *Transition of students with disabilities to postsecondary education: A guide for high schoolteachers.* Washington, DC: U.S. Department of Education. Retrieved from http://www2.ed.gov/print/about/offices/list/ocr/ ransitionguide. html

U. S. Department of Education (2014). Building the legacy: IDEA 2004. Retrieved from http://idea.ed.gov/explore/view/.

Wedler, H. B., Boyes, L., Davis, R. L., Flynn, D., Franz, A., Hamann, C. S., & Wang, S. C. (2013). Nobody can see atoms: Science camps highlighting approaches for making chemistry accessible to blind and visually impaired students. *Journal of Chemical Education.*

The White House (2013). *Education.* Retrieved from http:// www.whitehouse.gov/issues.

Wiazowski, J. (2009). Continuum of assistive technology for the blind and visually impaired. In Gierarch, J. (Ed.), *Assessing Students' Needs for Assistive Technology (ASNAT) Manual 5th Edition.* www.WATI.org/Resources.

World Health Organization. (2013). *Visual impairments and blindness.* Retrieved from http://www.who.int/mediacentre/factsheets/fs282/en/.

In: S.T.E.M. Education
Editor: Satasha L. Green

ISBN: 978-1-63117-806-1

Chapter 11

WORKING WITH ENGLISH LANGUAGE LEARNERS WITH SPECIAL NEEDS IN STEM

***Diane Torres-Velásquez*[1], *Deborah Roberts-Harris*[1], *Carlos López Leiva*[1], *Carol Westby*[2], *Gilberto Lobo*[3], *Barbara Dray*[4], *Ana Genoveva Martínez de la Cueva Astigarraga*[5] *and Jean Rockford Aguilar-Valdez*[6]**

[1]The University of New Mexico, NM, US
[2]Bilingual Multicultural Services, Inc., Albuquerque, NM, US
[3]Albuquerque Public Schools, Albuquerque, NM, US
[4]University of Colorado-Denver, CO, US
[5]Universidad Autónoma del Estado de Morelos, Morelos, Mexico
[6]St. Olaf College, MN, US

ABSTRACT

This chapter describes the need to implement equitable, instructional approaches to differentiate STEM education for ELL students with special needs. Based on frameworks such as Ganma, Nepantla, and Multiliteracies, we suggest a model that we call, *Complex Integrated Curriculum*. This student-centered model presents a differentiated STEM curricular approach that emphasizes learning and owning of knowledge processes that are parallel to the development of students' STEM identities (as students and as scientists). This curricular model carefully considers a responsive and adequate differentiation of the content, the process or activities, and the products that support the teaching and learning of ELL students with disabilities in STEM fields.

INTRODUCTION

Science classrooms can be rich language and science learning environments for English language learners (Lee, Quinn, & Valdés, 2013). Science also holds great potential for providing both cognitive and language development for students with special needs. When

English language learners (ELLs) have special needs, there are factors to consider in meeting those needs. In this chapter, we will briefly examine the prevalence of ELLs with special needs in the science classroom. We will present a case for the importance of Science, Technology, Engineering, and Mathematics (STEM) for this population. With the nationwide implementation of the Common Core State Standards (CCSS), the Framework for K-12 Science Education: Practices, Crosscutting Concepts and Core Ideas, and the next Generation Science Standards (NGSS), it will be important for teachers to not only be aware of the need for STEM instruction for all students, but also be prepared with pedagogical approaches, promising strategies and resources, and ways to differentiate STEM instruction. We present alternative frameworks as ways of supporting this student population in STEM education.

A Brief Summary of the Prevalence of ELLs with Special Needs

In United States schools, 13 percent (6.4 million) of all public school students are receiving special education services (Aud et al., 2013). A great number of these students come from culturally and linguistically diverse (CLD) backgrounds. CLD students are more likely to be disproportionately represented in special education with specific types of disability. For example, Native American students tend to be disproportionately identified as having specific learning disabilities. African American students are disproportionately labeled with emotional disturbances and intellectual disabilities, White students with autism, and Latino students with speech or language impairments (Ahram, Fergus, & Noguera, 2011; Donovan & Cross, 2002; Losen & Orfield, 2002; O'Connor & Fernandez, 2006; Skiba, et al., 2006). Three common patterns have been mentioned as possible factors affecting this overrepresentation:

"(a) A higher percentage of ethnic minority children than White children are assigned to special education;
(b) within special education, White children are assigned to less restrictive programs than are their ethnic minority counterparts; and
(c) the data—driven by inconsistent methods of diagnosis, treatment, and funding—make the overall system difficult to describe or change" (McDermott, Goldman, & Varenne, 2006, p. 12).

Additionally, some school systems on the basis of students' English skills (or lack of English proficiency) assign CLD students to special education programs (Artiles & Trent, 1994; Minow, 2001). Cultural deficit thinking affects how low-income and minority students' academic and behavioral discrepancies may be pathologized. There is also the pervasive belief that poverty influences cognitive ability. The fact that student race, class, and language are often predictors of referral to special education as a remedy of the perceived cognitive or behavioral deficits suggests that subjective judgments on diversity may be a factor in these bias perceptions (Ahram et al., 2011). Cultural differences are assessed as odd ways of behaving; CLD students' behaviors may be interpreted by teachers as 'disrespectful'; still, within this context some teachers are often reticent to discuss issues related to race, culture, and language (Skiba et al., 2006). Thus, "it is schools and not poverty that place minority students at heightened risk for special education placement" (O'Connor & Fernandez, 2006, p. 10). In this paper, we focus on English language learners who are receiving special education services and have an identified disability and who are in the Science, Technology, Engineering, and Mathematics (STEM) classes. While special needs may also refer to gifted students, we have focused in this chapter on ELL students with disabilities. In the United States, 10 percent of students in the public schools are ELL (Aud et al., 2013). ELLs represent a group of students with the most significant increase of numbers in U.S. schools. According to the U.S. Department of Education (2011), during the 10-year period from 1999 to 2009, the ELL population increased by 51% in some parts of the country, while the general PreK–12 student enrollment grew by only 7%. During 2006, of the students who were labeled as being limited English proficient and were between ages 6-21, 50.8% participated more than 80% of school day in a regular classroom, 26 % received 40-79% of their public school education in a regular classroom, and 20.9 % participated less than 40 % of their school time in a regular classroom (U.S. Department of Education, 2011). Thus, in 2006, almost half of all ELLs spent over 20% of their time at school out of a general education classroom. When you consider that students with special needs are drawn from the classroom for special education services, it is clear that students with special needs who are ELL spend an even greater amount of time outside of the general education classroom.

Table 1. Science Achievement Levels for 8th-Grade Students by Disability and English Language Learner Status, and Race/ethnicity (adapted table from NCES online Nation's Report Card)

Science Achievement Levels	School Services		English Proficiency		8th Grade
	Special Education	General Education	ELL	General Education	General Performance
Below Basic	*66%	31%	*83%	32%	35%
≥ Basic	34%	69%	17%	68%	65%
≥ Proficient	11%	34%	3%	34%	32%
≥ Advance	*0%	2%	*0%	2%	2%

By Ethnicity	Black 8th grade	American Indian 8th grade	Hispanic 8th grade	White 8th grade
Below Basic	*63%	*49%	*52%	20%
≥ Basic	37%	51%	48%	80%
≥ Proficient	10%	20%	16%	43%
≥ Advance	*0%	1%	*0%	2%

Note: We marked with asterisks those scores that we consider alarming. The National Center for Education Statistics (NCES) defines the different levels of achievement to describe student performance. NCES defines these levels as follows: *Basic* denotes partial mastery of prerequisite knowledge and skills that are fundamental for proficient work at each grade; *Proficient* represents solid academic performance. Students reaching this level have demonstrated competency over challenging subject matter; *Advanced* represents superior performance (NCES, 2012).

Table 2. Mathematics Achievement Levels for 4th- and 8th-Grade Students by Disability and English Language Learner Status, and Race/ethnicity (adapted table from NCES online data)

Mathematics Achievement Levels in 4th Grade	School Services		English Proficiency		4th Grade Math
	Special Education	General Education	ELL	General Education	General Performance
Below Basic	*45%	14%	*41%	15%	18%
≥ Basic	55%	86%	59%	85%	82%
≥ Proficient	17%	42%	14%	44%	40%
≥ Advance	2%	7%	1%	7%	7%
Mathematics Achievement Levels in 8th Grade	**School Services**		**English Proficiency**		**8th Grade Math**
	Special Education	General Education	ELL	General Education	General Performance
Below Basic	*64%	22%	*71%	24%	27%
≥ Basic	36%	78%	29%	76%	73%
≥ Proficient	9%	38%	5%	36%	35%
≥ Advance	2%	9%	1%	9%	8%

Levels by Race	Black		American Indian		Hispanic		White	
	4th grade	8th grade	4th grade	8th grade	4th grade	8th grade	4th grade	8th grade
Below Basic	34%	*49%	34%	*45%	28%	*39%	9%	16%
≥ Basic	66%	51%	66 %	55%	72%	61%	91%	84%
≥ Proficient	17%	13%	22%	17%	24%	20%	52%	44%
≥ Advance	1%	2%	2%	3%	2%	3%	9%	11%

Note: We marked with asterisks those scores that we consider alarming. The National Center for Education Statistics (NCES) defines the different levels of achievement to describe student performance. NCES defines these levels as follows: *Basic* denotes partial mastery of prerequisite knowledge and skills that are fundamental for proficient work at each grade; *Proficient* represents solid academic performance. Students reaching this level have demonstrated competency over challenging subject matter; *Advanced* represents superior performance (NCES, 2011, p. 7).

STEM academic performance data for ELLs and for students with disabilities is available through the National Center for Education Statistics' (NCES) online report (http://nationsreportcard.gov/). These data include average scores, achievement levels, and percentiles on the national levels of achievement collected through the National Assessment of Educational Progress. Focusing on achievement levels, we analyzed the NCES' data presented online and developed Table 1 (science achievement) and Table 2 (mathematics achievement).

Reports on academic achievement in mathematics and science portray an environment in which a significant percentage, at least 40%, of diverse student populations is performing Below Basic level in these two subjects. The gap between scores for White students and students in other student subgroups is also alarming. In 8th grade science, in comparison to

White students' average score, the gap is huge. For Blacks the gap is 34%, for Hispanics 26%, American Indians 22%. ELLs have a 48% gap from students who are not ELLs. Students with disabilities score 31% lower than students who have not been identified with a disability. In 8th grade mathematics, in comparison to White students' average score, Blacks score 31% lower than Whites, Hispanics 23% lower, American Indians 28% lower. ELLs score 42 points lower than students who are not ELL. Students with disabilities score 38 points lower than students who have not been identified with a disability. In both Science and Mathematics education, it is clear that we need to invest greater efforts to support diverse students in our schools. These results confirm what previous studies have pointed out: services provided to ELLs and diverse students do not always meet their academic needs (Artiles & Ortiz, 2002; Orosco & Klingner, 2010), specifically in STEM (Sotomayor, 2013). While we were able to find data indicating achievement for each subgroup including ethnicity, ELL, and disability, there are no subgroup data for students who are ELLs with special needs, making their needs almost invisible. We suspect that statistics for ELLs with special needs would be even more dismal, if they were available. Thus, it is crucial that we make all students' potential and needs visible in our STEM classrooms. Moreover, we believe that rather than focus on the numerical gap in student achievement, we should instead critically evaluate how we are working with all of our students and willingly appropriate and develop equitable approaches to support ELL students with special needs in STEM education.

EDUCATION EQUITY INCLUDES STEM EDUCATION

STEM education is important for all students, and not just for the "best and the brightest." Most often, we have looked at STEM education as something only for students who we expect to go on to higher education, and that usually excludes non-mainstream students. We need to recognize that ALL students can succeed and contribute when presented with opportunities to learn. Many of us, who are teachers, can point to students who have learning challenges, or who have linguistic challenges and still understand STEM well. Unfortunately, we seem to traditionally look at STEM education and these students as not part of the STEM equation. There is a pervasive societal norm of thinking that only certain individuals possess the capacity to learn STEM content. In fact, a report by the Carnegie-IAS Commission on Mathematics and Science Education, (2009, p.25) recognized there needs to be "clear recognition that math and science can be learned by all, although one in four hold doubts."

Classifying, categorizing and limiting students because of needs, ethnicity, race, gender, language, or economic issues has never been a viable or just option. Research has shown that limiting students in STEM education because of special needs, ethnicity, race, gender, language, or economic issues is not equitable or logical. For example,

> *Taking Science to School* (2007) and its companion reports for practitioners *Ready Set Science* (2008), *Learning Science in Informal Environments* (2009), and most notably *A Framework for K-12 Science Education* (2012) consistently highlight the point that provided with equitable learning opportunities, students from diverse backgrounds are

> capable of engaging in scientific practices and meaning-making in both science classrooms and informal settings (NRC, 2013, p.1).

U.S. law is very clear regarding the civil rights of persons with disabilities, immigrants, and students who speak a language other than English, in and outside of the public education system. The Americans with Disabilities Act (ADA), the Individuals with Disabilities Educational Act (IDEA), and No Child Left Behind (NCLB), as well as numerous court decisions, all are very clear: neither ethnicity, language, socioeconomic status, legal status as an immigrant, nor disability is an acceptable barrier to a child's access to education and educational success (Cook, n. d.).

We understand that our linguistically diverse students and students with special needs are capable of learning and contributing to STEM fields. All students should be offered access to opportunities in STEM areas of education at every grade level and should be encouraged to pursue careers in STEM areas. Michelle Cahill, Vice President of the Carnegie Corporation, and director of the Foundation's Urban Education Program was asked in an interview on C-Span (2009) why it is critical for stronger math and science education to be available for all students. She responded, "We need a country that's unified," she said, "and in which there is opportunity for all."

Opportunities in Innovation

STEM fields are the foundation for innovation, and STEM education provides opportunities for students to develop innovative products and solutions. The National Center for Technology Innovation and Center for Implementing Technology in Education asserts,

> Science, technology, engineering, and math (STEM) education serves as the foundation of innovation in our society. Innovative products often derive from a problem or challenge that requires a unique solution, making it imperative that all students, including students with disabilities, have access to a rigorous STEM curriculum (CITEd, 2009). (no page)

Often students who are able to think outside the box, or have a unique approach to learning or understanding, or a perspective that is broader than the norm are the students who can develop novel, new ideas or methods. Certainly students who face challenges in learning have often acquired specialized strategies to succeed, which could possibly be applied to the development of innovative products and or solutions to problems.

New Global Workforce

As the need for scientists and engineers increases, we cannot continue to ignore or exclude students due to language or disability. The Committee on Equal Opportunities in Science and Engineering (2006) believes that students with disabilities comprise one of the largest untapped pools of potential scientists, engineers, technologists, mathematicians, and technicians. Our current economy and society requires an increasingly rigorous knowledge base. Level the Playing Field Institute states, "education at all levels in science, technology, engineering, and mathematics—STEM—develops, preserves, and disseminates knowledge and skills that convey personal, economic, and social benefits. (Level the Playing Field, n. d.)." According to the National Council of Governors, The global economy has "flattened" the world in terms of skills and technology.

> A new workforce of problem-solvers, innovators, and inventors who are self-reliant and able to think logically is one of the critical foundations that drive innovative capacity in a state. The K–12 (kindergarten through grade 12) education system, with the support of postsecondary education, the business sector, foundations, and government, must ensure that (1) *all* (emphasis added) students graduate from high school with STEM competencies to become this work-force; and (2) a greater number of students graduate from high school as potential professionals in STEM fields (Thomasian, 2011, p.9).

Level the Playing Field Institute is an organization whose mission is to find ways to eliminate the barriers in STEM education and careers experienced by people of color who are underrepresented in STEM and to promote talented individuals to help advance our nation. They believe that everyone should have equal access to opportunities, and that in order to "level the playing field" education is the place to start. They believe STEM education is important for all, but especially for students who have historically been disenfranchised.

Appreciation of the Natural World

In an effort to help all students learn to appreciate nature, the Department of the Interior (DOI) has set aside funding for learning opportunities to engage students, especially under-served students, to learn about and appreciate the outdoors and natural settings. They found that young people today from all walks of life spend much less time interacting in natural settings due to increased interaction with technology, that minority populations spend less time outdoors than their peers, and that this decreased exposure leads to less familiarity, understanding, and appreciation of nature (U.S. Department of Interior, 2013). Although the focus of much of their work is primarily at the middle school level, their intention is to engage learners of all ages to nurture a "sense of wonder about the natural world that will inspire future learning (p.13)." They recognize that all children at a very young age can experience and enjoy the natural world.

Making Informed Decisions about our World

One focus of the DOI is to see all students succeed and to ensure that all students, as future decision makers, are engaged in science related fields. Their STEM education and employment Five-Year Goal articulates the present-day need to teach advocacy and protection of our environment:

> That our youth and the American public become scientifically literate stewards of our natural and cultural heritage and that today's youth, especially those under-represented in STEM fields of study, become inspired to choose career paths at DOI or related agencies and partners (p.2).

Additionally, their mission statement claims that they are working to create an informed citizenry by advancing "the science, engineering, and technology that inform natural and cultural resource management, natural hazards response, and decision-making on critical issues that impact our nation, the world, and society (NRC, 2013, p. 2)."

Experiences that provide students with the understandings and practices needed to make decisions based on scientific knowledge will protect our lives, our health, and our environment. Not including ELL students with special needs would be in direct conflict with U.S. law, and should create a conflict for us as educators with personal and professionally ethical commitments to advocate for all students and their families and to remove all barriers to their success (Cook n. d.). Authors of a report by the Commission on Mathematics and Science Education (2009) noted:

> As a society, we must commit ourselves to the proposition that all students can achieve at high levels in math and science, that we need them to do so for their own futures and for the future of the country, and that we owe it to them to structure and staff our educational system accordingly (p.9).

How do we move in the direction of motivating, empowering and inspiring our students with linguistic and learning challenges into the STEM world? We can be intentional in our instruction and promote the idea that continued education is expected for all of our students because they can meet our high expectations. We must be at the forefront of implementing practices that continue to confront discrimination and bias that exist at all levels of education. Beyond K-12, we need to be part of the bridge that closes the gaps that exist in college admissions, retention and successful graduation in STEM areas for our ELL students with special needs. We can encourage our students to envision themselves as able, and to see themselves as having the social and cultural capital to continue their education in STEM fields. Inviting students to embrace their multiple identities and take on the challenges they will face is not only our obligation, but also an opportunity to improve the world.

One model that captures the dynamics and complexity of teaching special needs students who are ELLs is the International Classification of Functioning, Disability, and Health (World Health Organization, 2001). Rather than defining a person merely by disability or with a categorical label, this model acknowledges an individual's level of functioning across contexts (see Figure 1).

The concentric circles of this model frame the levels of influence. The arrows indicate the possible interactions between levels. In this model it is recognized that a student brings to the classroom all of who they are, including factors internal and unique to them, and societal factors that are external to them but affect their daily lives. These internal and external factors interact to influence who they are and how they function in every aspect of their lives. Individual factors that shape identity could include culture, language(s), age, education, prior experiences, self-esteem, identity, trauma, personality, level of care, challenges and strengths in how the physical, psychological, and cognitive functions of the body, and social structures that define the individual's ability to interact in family, community and school activities. The model does not assume that individuals function in isolation. There is a consideration for external factors such as societal attitudes, historical traumas, policies, etc., that affect the student, his or her family life and ultimately the student's learning. For students with special needs who are ELLs, policies that cover rights or boundaries in areas of immigration, tribal sovereignty, language use, family engagement, disability, and many more are just a few examples of external factors that may affect the student's education, family life, and school curriculum, and thus affect the individual's ability to fully participate in the world. This integrative model takes into account all supports and restrictions across all domains including all areas of family, school, and community. This model provides space for the interaction of family beliefs and practices, the classroom curriculum, and the barriers faced by a student's particular levels of functioning.

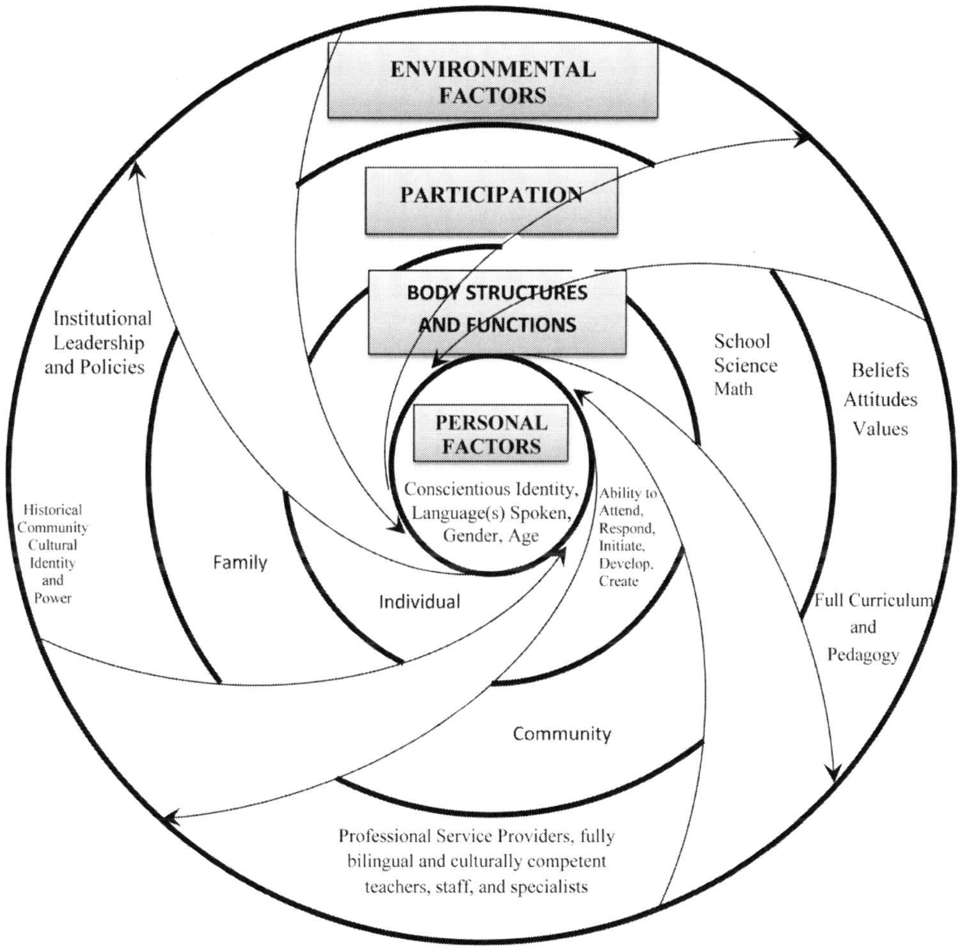

Figure 1. The International Classification of Functioning, Disability and Health (ICF) and Science Education for English Language Learners with Special Needs. Adapted from: Westby and Inglebret (2012); CHiXapkaid and Inglebret (2007); Inglebret, Brownfield, and CHiXapkaid (2009); World Health Organization (2002).

Westby & Inglebret (2012) used the weaving of the mat to represent the factors, tensions, and structures that vary and interconnect with nested circles representing individuals, traditional, and complex systems of knowledge and language, while giving us the opportunity to draw upon the wealth of the home culture and using indigenous knowledge as an asset. We modified this model to reflect STEM education.

Pedagogical Approaches to STEM for English Language Learners with Special Needs

Some pedagogical approaches that have been deemed productive in STEM fields are currently being supported by National Standards, such as CCSSI (2010) and Next Generation Science Standards (2013). In science education, for example, current pedagogic perspectives support an active, constructive, experiential, and collaborative learning process (Bereiteer & Scandamalia, 2003). Learning and cognitive self-regulation are supported through meaning-rich contexts with challenging, realistic, and open inquiry processes (Warren & Rosebery, 2008) in which demonstrations and continuous use of the discourse with learners themselves are essential to understanding and using science practices (Roth & Lee, 2002). Next Generation Science Standards (2013) promote practices necessary for doing science. These practices are: (a) asking questions and defining problems, (b) developing and using models, (c) planning and carrying out investigations, (d) analyzing and interpreting data, (e) using mathematics and computational thinking, (f) constructing explanations and designing solutions, (g) engaging in argument from evidence, (h) obtaining, evaluating, and communicating information. These practices for effective science teaching are consistent with the findings reported in the study described below by NCES. Students in general education whose teachers used hands-on projects every day or almost every day scored higher than students whose teachers used hands-on projects in class less frequently. Similarly, teachers who reported that their students worked together on science projects weekly or daily scored higher than students whose teachers reported that students did so monthly or never. Finally, students who did related science activities but not directly connected to schoolwork also scored higher than students who only did school work (NCES, 2013).

In mathematics education, emphasis has been made on a problem solving approach (Schoenfeld, 1985; NCTM, 2000) where students learn conceptual and procedural knowledge by constructing and applying mathematical knowledge. Students' previous knowledge and understandings are used as a platform on which to start building new knowledge (Van de Walle, Karp, & Bay-Williams, 2013) in a communicative environment that supports meaning making and development of mathematical discourse (Chapin, O'Connor, & Anderson, 2009). The National Common Core State Standards for Mathematics (CCSS-M, 2010) promote mathematical practices related to this problem solving and active way of learning mathematics. The mathematical practices teach students to: (1) make sense of problems and persevere in solving them, (2) reason abstractly and quantitatively, (3) construct viable arguments and critique the reasoning of others, (4) model the practice of problem solving with mathematics, (5) use appropriate tools strategically, (6) attend to precision, (7) look for and make use of structure, and (8) look for and express regularity in repeated reasoning. For ELLs and students with special needs, *problem solving* not only promotes the application of concepts, it can also be part of everyday life. Certainly, children come to school with a good deal of mathematical knowledge, even students with special needs (Baroody, 1987). They have natural strategies to solve certain problems and do not need to be taught these strategies. The task for teachers is using, refining, and enhancing this informal knowledge, giving students opportunities and encouragement to model the actions and or relationships in a problem (Baroody, 1987; Carpenter & Moser, 1982; Van Garderen, 2012).

More recently, in mathematics education for students with special needs, the most significant strategies for improving mathematics of students with disabilities include: (a) teaching students the use of heuristics to solve problems, and (b) using explicit instruction to solve specific problems by applying specific strategies (Gersten et al., 2009). Specifically to students with disabilities who are ELLs, it has been suggested that teachers pay attention to interactions they have with family members and peers outside of class, as well as among student peers in the classroom, so that these interactions may become resources for learning and problem solving. Communications may take place around problems that support students' reasoning in activities which are meaningful to students, with mathematical problems or tasks linked by the teacher to their students' lives, knowledge, and interests (Torres-Velásquez & Lobo, 2004/2005). Additionally, for students with disabilities in mathematics, providing feedback and using graphic representations are strategies that have been proven useful (Van Garderen, 2012; Van Garderen & Montague, 2003). The representations are helpful to illustrate procedures and mathematical relations, and are a helpful tool for moving concepts from concrete to abstract. Mathematics and science become more relevant within an environment where students are encouraged to think aloud and write their explanations and ideas using the language they know best, often their native language. As they learn to do mathematics and science, it is important for ELLs with disabilities to learn the more abstract academic language of both mathematics and science in both their native language and English. This allows students to work in the language of their choice, and also allows them to communicate with their families at home using what they have learned in the classroom. An important job for the teacher—as suggested by Collier (2004)—is to analyze and think ahead about how students engage with mathematics problems and their use and knowledge of the language(s) and accordingly sequence directions in an order that will promote scaffolding to understand, learn, and transfer or generalize mathematical and language skills and knowledge.

Ganma, Nepantla, and Multiliteracy Frameworks

The families of ELLs may come from other countries, or they may be indigenous to this country. When you consider that Spanish speakers, for example, may come from any country where Spanish is spoken and, just as with English, language varies by region and dialect, it becomes clear that students who speak the same language may not identify with the same culture, or even race. Increasingly in today's society, families are multiethnic (McCubbin et al., 2010). In North America and on other continents, we are just beginning to acknowledge the influence of Black ancestors in indigenous communities. In various tribes in New Mexico, we have tribal members who are Black. From the Dominican Republic and islands of the Caribbean, Afro-Latinos come to the states completely fluent in Spanish because that is their native language. Especially for teachers, it is important to recognize that race is not a predictor of culture or language.

Learning about our students' identities can help teachers be most effective in working with ELL students in our STEM classrooms. Students may come with families who hold traditional beliefs, values, and styles of living of their home culture and prior generations; or students may have families who are developing bicultural identities, where they are holding onto some traditional beliefs and practices, while adopting the lifestyle of a second culture or parts of the dominant culture. It is possible that an ELL student with special needs enters our

STEM classroom already speaking two languages; say Spanish and Nahuatl, before even moving to the United States, where they learn to speak English as their third language. Maintaining the home language is critical so that students do not experience language loss and its cognitive and social consequences (Fillmore, 1991). Some students who have worked to establish an identity in the dominant culture may experience a pan-renaissance, where they work to redefine and reconfirm their heritage culture, language and lifestyle (Westby & Inglebret, 2012). A student's identity may influence everything at school from how they interpret the curriculum to who they tend to seek as friends or mentors.

Ganma, an Australian aboriginal concept, is considered to be the place where knowledge is created. The concept provides a metaphor for ways that multiple cultures come together in a positive and constructive way. Elders teach that saltwater from the sea mixing with freshwaters from rivers forces both streams to combine, creating foam. This foam signifies new understandings, and the mixing of waters below the surface signifies deeper understandings. It is a beautiful way to envision two cultures and their languages coming together and creating knowledge; for example Western culture and knowledge mixing with Aboriginal culture and knowledge. One culture does not dominate the other; instead they mutually engulf each other, flowing into a common lagoon (Pyrch & Castillo, 2001). The primacy of all groups' backgrounds and experiences are valued and a new culture is created (Westby & Inglebret, 2012). If the waters can stay in balance, the combining nutrients nourish diverse forms of life (Cazden, 2000).

In this chapter, we introduce a Multiliteracies Framework as a tool for implementing the ICF model for assessment and instructional purposes with special needs students who are ELLs (Healy, 2008; Westby, 2010). This framework can serve as an interface between two cultures where new types of knowledge can be generated. It requires that participants, in this case families, teachers, and learners, become familiar with their roles in a dynamic action-filled environment (Martin, 2008). In this framework, literacy takes on a deeper and richer meaning than simply decoding and comprehending written words, phrases and paragraphs. Multiliteracy involves reading, comprehending, and interpreting all information across multiple modes of communication in the environment, across technology and multiple contexts. Imagine literacy in science education that includes reading nature and the natural environment, oral histories and traditions, gestures, videos, telecommunication, spatial and spiritual linguistic modes across cultures, in addition to graphic and written communications.

This framework allows students and teachers to cross cultural borders in the curriculum. English language learners with special needs enter, straddle, and negotiate multiple cultures. Anzaldúan Theory (Anzaldúa, 2002) is one that proposes a space between the home and dominant culture where a student negotiates multiple worlds. In Aztec mythology, this in-between space is called Nepantla. This crossroads is the place where students have room to integrate the parts of their self that come from the multiple cultures to which they belong, and to negotiate their identity as students of science (Aguilar-Valdez, et.al, 2013). Nepantla, the in-between state, provides a place to embody the concept of Ganma, valuing both the home culture and the dominant culture, as students learn science in and from both worlds, learn to confront misconceptions, and integrate diverse values, beliefs and traditions.

There are four components that are used to implement Multiliteracies Pedagogy, which include: (a) situated practices, (b) overt instruction, (c) critical framing, and (d) transformed practice (Healy, 2008; New London Group, 1996). Situated practice embeds and grounds learning in the experiences and knowledge that has been established at home, in the home culture, and in the community. Overt instruction is the explicit teaching of concrete and specific skills or knowledge and may come from the knowledge base of the school and standardized curriculum. Especially for ELLs with special needs, this knowledge may be different from that of the home culture and language. Critical framing prepares students to analyze and question the world around them. Student learning is framed as something they own and they are allowed to question assumptions. Students learn to own their learning as they apply it to real-life problems or issues. They problem solve and create doable solutions to make their world a better place using transformed practice. Figure 2 shows an example of how the four components may interact in an environmental education curriculum situated in a dual-language Spanish-English STEM sixth grade class in the Southwestern United States.

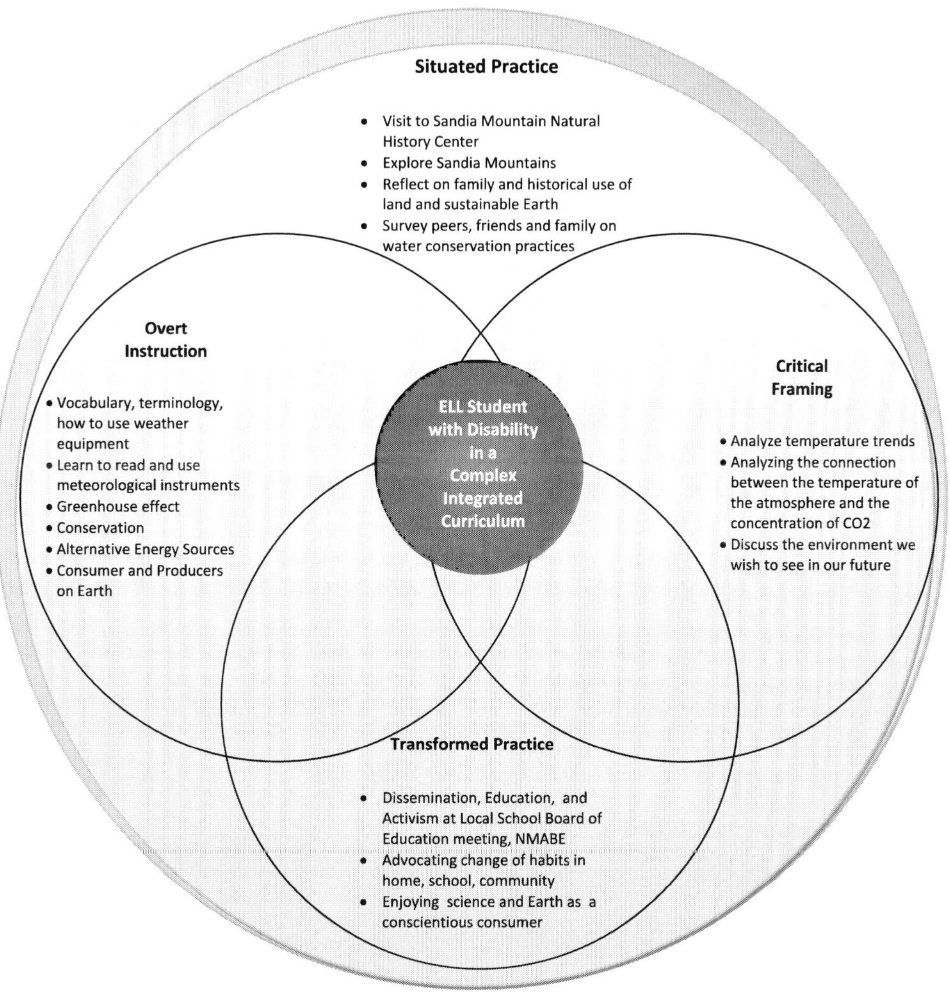

Figure 2. Complex Integrated Curriculum: Multiliteracy Framework in a Dual-Language STEM 6th grade classroom studying Environmental Education.

Using the Multiliteracies Pedagogy and an Environmental Education Curriculum developed by Gilberto Lobo and his instructional team, students in a sixth grade dual language science classroom learned to become meteorologists, advocate for conservation, and be conscientious and educated consumers. During the summer of 2012, the dual language 6th grade teachers attended the Sandia Mountain Natural History Center where they participated in the Environmental Curriculum Workshops designed for middle school teachers. Teachers developed their curriculum based on what they knew about their students and families. For the students, the curriculum began in fall of 2012 with a visit to the Sandia Mountains. All activities were situated in the students' prior knowledge and experiences, or in shared experiences at the school, in the community, or in the mountains. This Situated Practice gave students a sense of the science they were learning, and a living learning environment in which to share vocabulary and knowledge they were explicitly taught. Situated Practice is especially important for ELLs with special needs. The meaningful and concrete setting, activities, and language provide a layer of support that encourage understanding and application of new knowledge. All content was taught in English and Spanish. Because of the importance of Situated Practice in all aspects of this curriculum, Figure 2 is designed to convey that all learning in this curriculum occurred in the situated context. Students were overtly taught the vocabulary and tools of meteorologists. They learned factual information about the mountain they see daily from afar. Students learned to read and use the tools of meteorologists, and learned to collect data to address the issues of conservation, and Climate Change. Teachers differentiated instruction in all areas based on students' strengths and needs, particularly in the explicit teaching of facts and concepts from the scientific, technological, and mathematical domains used by real scientists in the science of Meteorology. Please see specific overt teaching strategies included in the section that follows.

A unit on Environmental Education easily lent itself to Critical Framing as students not only learned to analyze temperature trends, but also to visualize Earth as they would like to see for themselves as adults and for their future generations. Students delivered a well-informed presentation at their district school board meeting and at the New Mexico Association for Bilingual Education (NMABE) Annual Conference to demonstrate what they had learned. As part of their presentation, they used the mathematics and science languages of both Spanish (their home language) and English. They shared alarming news about Climate Change, pointing out data they had collected as scientists while measuring water usage in home and in community, and they challenged audience participants to consider what steps they might commit to save our environment and slow the rate of Climate Change. This example of Transformed Practice helped students and audience members create a better world.

In this example, teachers developed a curriculum for a relevant topic in STEM, using a Multiliteracies Framework and honored the functioning, abilities, and health of each student. They helped students navigate the multiple worlds of school, community and environment as students, community leaders, and scientists.

Differentiating STEM Instruction for ELL Students with Special Needs

Differentiated instruction is an approach to teaching and learning for students of differing abilities in the same class. The intent of differentiated instruction is to maximize each student's growth and individual success by meeting each student where he or she is, and assisting in the learning process. Differentiated instruction is essential for ELL students with special needs. It involves creating multiple paths so that students of different abilities are able to access the academic curriculum (Tomlinson, 2001). Educators can assist in differentiating instruction for ELL students by:

- identifying what is required in the learning task, such as vocabulary knowledge, syntactic understanding, the ability to make inferences, the ability to work independently or in a group
- determining students' language, cognitive, and social strengths and weaknesses
- designing lesson goals that address both common core standards and the needs of individual students
- developing differentiated lesson objectives (based on common core standards) for what should be learned for all students, some students, a few students
- employing a variety of teaching strategies in differentiated instruction

Instruction can be differentiated in terms of the content that students are to learn, the processes used in teaching the content, and the product that students are to produce to demonstrate their learning.

Differentiating Content

What information should the students learn? In any particular academic content unit, what is the most important information for students to learn? What are the key points that the teacher would like everyone to understand? What additional content could most of the students in the class learn? What content might only a few of the high-functioning or gifted students learn (Schumm, Vaughn, & Leavel, 1994; Watson & Houtz, 2002)? Figure 3 shows differentiated content for an elementary school unit on geology. In considering the content to be taught, the teacher must determine the underlying vocabulary and sentence patterns that the students will need in order to access the content.

In the explicit teaching of language, providing ELLs with language frames to use when discussing STEM concepts can reduce complexity of the task and increase comprehension of mathematical and scientific concepts. Language frames are partially constructed cloze statements that highlight the academic language and syntax required to communicate ideas. They provide students practice in the discourse patterns expected for the subject matter. Rather than struggling both to retrieve the vocabulary and concepts and organize this information into coherent sentences, students use the frames to facilitate their ability to explain the concepts (Donelly & Roe, 2010; Fisher, Frey, & Ross, 2009). Language frames are effective tools, but students require teacher modeling of these tools to build their capacity for holding meaningful discussions. Table 3 shows sentence frames of increasing complexity that can be used for comparing and contrasting concepts.

Discussions in science classes often require the use of argumentation, a structurally complex discourse style that involves offering and responding to claims, providing and asking for evidence or justifications, and analyzing those claims to formulate a decision, to formulate arguments and counterclaims.

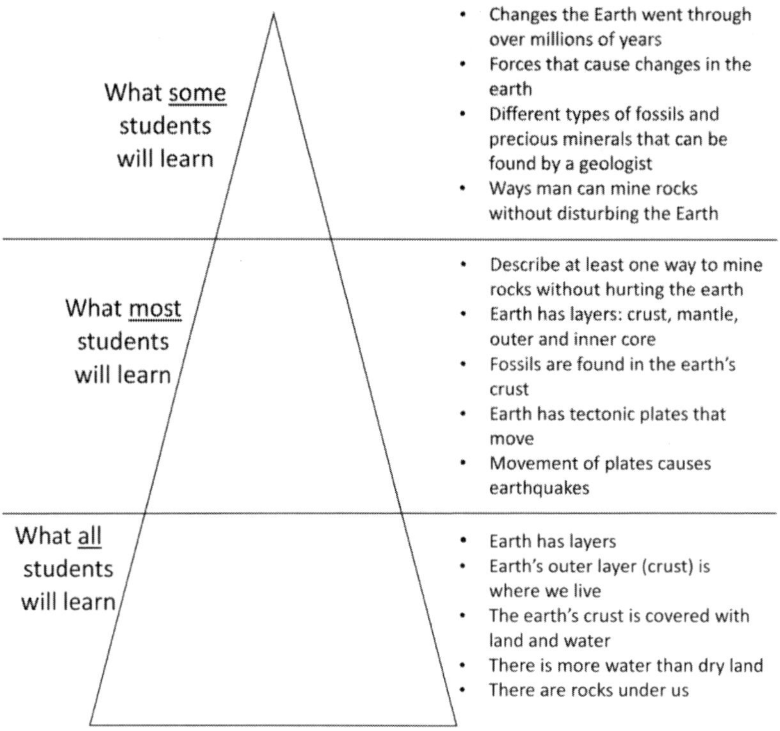

Figure 3. Differentiated content for geology unit.

Table 3. Sentence Frames for comparison

	Simple sentence	Comparative sentence	Complex comparative sentence
Sentence frame with vocabulary underlined	Eagles are large birds of prey. Kestrels are small birds of prey.	Eagles and kestrels are both birds of prey, but eagles are large and kestrels are small.	A main difference between eagles and kestrels is eagles are large birds of prey while kestrels are small birds of prey.
Sentence frame with vocabulary removed	____ are ____. ____ are ____.	_____ and _____ are both _____, but _____ are _____, and _____ are _____.	The main difference between _____ and _____ is _____ are _____, while _____ are _____.
Sentence frame with vocabulary underlined	Tornados are violent storms that begin on land. Hurricanes are violent storms that begin over the ocean	Tornados and hurricanes are both violent storms, but tornados begin over land and hurricanes begin over the ocean.	A main difference between tornados and hurricanes is tornados are violent storms that begin over land in the US Midwest while hurricanes are violent storms that begin over the ocean south of Florida.
Sentence frame with vocabulary removed	____ are ____ that ______. ____ are ____ that ____.	_____ and _____ are both _____, but _____ are _____ that ____, and _____ are _____.	The main difference between _____ and _____ is _____ are _____ that _____, while _____ are _____ that____.

The following are some of the types of language frames that students could learn to use in science arguments (Fisher, Frey, & Ross, 2009):

- Making a claim: I noticed ___, when _____.
- Providing evidence: I know that _____ is _____ because
- Requesting evidence: How did you know that ____ was ____?
- Offering a counterpoint: My idea is ______ because I read______.
- Inviting speculation: I wonder what would happen if _____.
- Reaching consensus: I agree with ____ because ____.

Differentiating Process or Activities

Differentiating process involves varying learning activities or strategies as a way to provide appropriate and more easily accessible methods for students to explore the concepts. Engaging ELL students in Instructional Conversation (IC) (Saunders & Goldenberg, 1999) is one way to differentiate the process of instruction. Rather than engaging in question-answer routines, when employing IC, teachers encourage expression of students' ideas, build upon information students provide and experiences they have had, and guide students to increasingly sophisticated levels of language and understanding. ICs contextualize abstract STEM concepts for students. They help students link their knowledge of concrete experiences to abstract STEM concepts by encouraging them to think critically, analyze, and synthesize ideas using language as a tool. Instructional Conversations are particularly important for ELL students, many of whom may not receive the opportunities for the development of concepts and academic language (the language of school and textbooks) in their native language since they are not attending school in that language and are not being schooled in their home country. Conceptual and linguistic development at home and school and in both languages is the goal.

Visual supports such as graphic organizers, maps, charts, and videos can also assist ELL students with special needs by reducing the load on verbal working memory. Partially completing graphic organizers or varying the complexity of the graphic organizers can effectively facilitate differing levels of cognitive processing for students of differing ability. Figure 4 shows a graphic organizer developed by ELL students with learning disabilities for a 7th grade science lesson on plant taxonomy. The benchmark and performance standard for the general education students were:

- Benchmark: Explain the diverse structures and functions of living things and the complex relationships between living things and their environment.
- Performance standard: Know how to classify organisms: domain kingdom, phylum, class, order. family, genus, species

For students with learning disabilities, the standard was modified to: Students will classify pictures of plants in their environment according to their characteristics. The students took photos of plants around the school and used the computer program *Inspiration* to display the relationships among the plants.

Figure 4. Graphic organizer for plant taxonomy.

The goal of all these activities is to promote understanding but more importantly is to make language (not only English) the central tool and authentic means for learning and developing a mathematical—and/or scientific—discourse. Throughout this process the interactions between students and teacher, and among students themselves, are embedded in a context, in an activity where language input and output matter and are expected and socially supported. This kind of culture of a STEM classroom community will mediate the development and appropriation of the target understandings and discourse practices (Chval & Khisty, 2009; Khisty & Chval, 2002).

In this context, teacher and students engage in a purposeful, mediated communication in which students have frequent opportunities and constructive feedback in participation, so the learning and teaching activity is centered beyond words and terminology, but on students' understandings and meaning making (Chval & Khisty, 2009); then the teachers and students create a language-based classroom culture for the learning of STEM concepts.

Differentiating Product

Students can be given multiple options for expressing what they know and understand. Do they do a drawing or produce an oral or written report? If written, is the report done by hand or on a computer. For the plant taxonomy unit, the general education students took a written test on the content. The students with learning disabilities were allowed to use the graphic organizer to write a summary of what they learned about plants.

What is the level of complexity required for the product [e.g., using Bloom's taxonomy, does the task require remembering, understanding, applying, analyzing, evaluating, and creating] (Anderson & Krathwohl, 2001). A general education fifth grade science teacher with a number of ELL students with special needs in her class did an ecology unit. Table 4 shows the range of evaluation (or product) activities she provided for her students based on Bloom's Taxonomy.

Table 4. Evaluation Activities Based on Bloom's Taxonomy of Objectives, Cognitive Domain

- Remembering: Draw a picture of your terrarium.
- Understanding: Describe the procedures you used to put your terrarium together
- Applying: Put worms and a variety of "garbage" into your terrarium. Record your observations over a week. Explain what you saw happen.
- Analyzing: Compare and contrast the lives of plants and animals in a terrarium
- Evaluating: A housing developer wants to buy a large section of the Bosque and build very expensive homes there. Write a letter to the editor of the newspaper about your opinion of this idea.
- Creating: Based on what you have learned about ecological systems, develop a plan for what the city of Albuquerque could do to preserve the Bosque (the wooded area near the river).

Complex Integrated Curriculum

In the Multiliteracies Framework, instructional strategies, such as the ones provided above, support Overt Instruction. When nestled in Situated Practice, Overt Instruction and Critical Framing come together for student learning, just as the waters in Ganma. Students learn to look critically at the environment, for example, while learning specific facts, knowledge and how to implement tools as a scientist. Transformed practice, using this new learning for positive actions, helps students apply and solidify their knowledge.

Framework components come together, just like the waters forming foam in Ganma, and blend into one stream of water. The languages and cultures of home and school are both valued and are both available for teaching and learning science. We suggest that this model helps students learn and own their knowledge, as they learn to be scientists. We place the student in the center of this model, within a curriculum that allows them to consider who they are as a student and as a scientist. We call this teaching Complex Integrated Curriculum, as an approach to understand disability and envision pedagogy using students' abilities across settings. We propose that this curriculum has great potential for conveying the wonderment of teaching and learning science and working with special needs students who are ELL.

REFERENCES

Aguilar-Valdez, J. R., LópezLeiva, C. A., Roberts-Harris, D., Torres-Velásquez, D., Lobo, G., & Westby, C. (2013). Ciencia en Nepantla: The journey of Nepantler@s in science learning and teaching. *Cultural Studies of Science Education.*

Anderson, L. W., & Krathwohl, D. R. (Eds.) (2001). *A taxonomy for learning, teaching, and assessing: A revision of bloom's taxonomy of educational objectives.* Boston, MA: Allyn & Bacon. Boston.

Anzaldúa, G. (2002). Now let us shift…the path of conocimiento…inner work, public acts. In G. Anzaldúa and A. Keating (Eds.), *This bridge we call home: Radical visions for transformation* (pp. 540-578). New York, NY: Routledge.

Ahram, R., Fergus, E., & Noguera, P. (2011). Addressing racial/ethnic disproportionality in special education: Case studies of suburban school districts. *Teachers College Record, 113*(10), 2233-2266.

Artiles, A. J., & Ortiz, A. A. (Eds.). (2002). *English language learners with special education needs: Identification, assessment, and instruction.* McHenry, IL: Delta Systems.

Artiles, A. J., & Trent, S. C. (1994). Overrepresentation of minority students in special education: A continuing debate. *The Journal of Special Education, 27*(4), 410-437.

Aud, S., Wilkinson-Flicker, S., Kristapovich, P., Rathbun, A., Wang, X., & Zhang, J. (2013). *The Condition of Education 2013* (NCES 2013-037). U.S. Department of Education, National Center for Education Statistics. Washington, DC.

Baroody, A. J. (1987). *Children's mathematical thinking: A developmental framework for preschool, primary, and special education teachers.* New York, NY: Teachers College Press.

Bereiteer, C., & Scandamalia, M. (2003). Learning to work creatively with knowledge. In E. De Corte, L. Verschaffel, N. Entwistle, and J. van Merriënboer (Eds.) *Powerful learning environments: Unraveling basic components and dimensions* (pp. 55-68). Oxford, UK: Pergamon, Elsevier Science Ltd.

Cahill, M. (2009). *Interview on C-Span.* Retrieved from: http://opportunityequation.org/news-press/michele-cahill-c-span

Carnegie Corporation of New York & Institute for Advanced Study Commission on Mathematics and Science Education. (2009). The opportunity equation. New York, NY.

Carpenter, T. P., & Moser, J. M. (1982). The development of addition and subtraction problem-solving skills. In T. P. Carpenter, J. M. Moser, & T. A. Romberg (Eds.), *Addition and subtraction: A cognitive perspective* (pp. 9-24). Hillsdale, NJ: Lawrence Erlbaum Associates.

Cazden, C. B. (2000). Four innovative programmes: A postscript from Alice Springs. In B. Cope & M. Kalantzis (Eds.), *Multi-literacies and the design of social futures* (pp. 321 - 332). London & New York: Routledge.

Chapin, S. H., O'Connor, C., & Anderson, N. C. (2009). *Classroom Discussions: Using Math Talk to Help Students Learn, Grades 1-6.* Sausalito, CA: Math Solutions Publications.

Chval, K. & Khisty, L. L. (2009). Bilingual Latino students, writing and mathematics: A case study of successful teaching and learning. In R. Barwell (Ed.), *Multilingualism Mathematics Classrooms: Global Perspectives* (pp. 128-144). Tonawanda, *NY*: Multilingual Matters.

Collier, C. (2004). Developing instructional plans and curricula for bilingual special education students. In L. M. Baca & H. T. Cervantes (Eds.), The bilingual special education interface (pp. 230-273). Upper Saddle River, NJ: Pearson Education Inc.

Cook, M. (n.d.) Language Difference v Language/Learning Disability: The art and science of distinguishing between diversity and disorder. Retrieved from: http://onlineresources.wnylc.net/pb/orcdocs/LARC_Resources/LEPTopics/ED/LnguageDifferenceLanguageLearning%20Disability.pdf

Committee on Equal Opportunities in Science and Engineering. "Biannual Annual Report to Congress" 2005-2006. Available at from the National Science Foundation, http://www.nsf.gov.

Common Core State Standards Initiative (CCSSI). (2010). *Common Core State Standards for Mathematics. Common Core State Standards (College- and Career-Readiness Standards and K–12 Standards in English Language Arts and Math).* Washington, D.C.: National Governors Association Center for Best Practices and the Council of Chief State School Officers.

Donelly, W.B., & Roe, C.J. (2010). Using sentence frames to develop academic vocabulary for English learners. *The Reading Teacher, 64*, 131-136.

Donovan, S., & Cross, C. (2002). *Minority students in special and gifted education.* Washington, DC: National Academy Press.

Fisher, D., Frey, N., & Ross, D. (2009). The art of argumentation: Fourth graders practice with language frames to learn the process of argumentation in inquiry-based instruction. *Science and Children*, 28-31.

Fillmore, L. W. (1991). When learning a second language means losing the first. *Early childhood research quarterly*, *6*(3), 323-346.

Gersten, R., Chard, D. J., Jayanthi, M., Baker, S. K., Morphy, P., & Flojo, J. (2009). Mathematics instruction for students with learning disabilities: A meta-analysis of instructional components. *Review of Educational Research, 79*(3), 1202-1242.

Healy, A. (2008). Indigenous worldviews, knowledge, and research: The development of an indigenous research paradigm. *Journal of Indigenous Voices in Social Work, 1,* 1-6.

Hemphill, F.C., & Vanneman, A. (2011). *Achievement gaps: How Hispanic and White students in public schools perform in mathematics and reading on the national assessment of educational progress* (NCES 2011-459). National Center for Education Statistics, Institute of Education Sciences, U.S. Department of Education. Washington, DC.

Khisty, L. L. & Chval, K. (2002). When teachers' talk matters. *Mathematics Education Research Journal*, 14(3), 154-168.

Lee, O., Quinn, H., & Valdés, G. (2013). Science and language for English language learners in relation to next generation science standards and with implications for common core state standards for English language arts and mathematics. *Educational Researcher*, *42*(4), 223-233.

Level the Playing Field Institute. (n.d.) Retrieved from: http://www.lpfi.org

Losen, D., & Olsen, G. (Eds.). (2002). *Racial inequity in special education*. Cambridge, MA: Harvard Education Press.

Martin, (2008). The intersection of Aboriginal knowledges, Aboriginal literacies, and new learning pedagogy for Aboriginal students. In A. Healy (Ed.), *Multiliteracies and diversity in education: New pedagogies for expanding landcapes* (pp.58-81). South Melbourne, Victoria, Australia: Oxford University Press.

McCubbin, H., Ontai, K., Kehl, L., et al. (2010). *Multiethnicity and multiethnic families: Development, identity and resilience*. Honolulu, HI: Le'a Publications.

McDermott, R., Goldman, S., & Varenne, H. (2006). The cultural work of learning disabilities. *Educational Researcher, 35*(6), pp. 12-17.

Minow, M. L. (2001). *Limited English proficient students and special education*. Wakefield, MA: National Center on Accessing the General Curriculum. Retrieved from http://aim.cast.org/learn/historyarchive/backgroundpapers/lep_sp_ed

National Center for Education Statistics. (2012). *The Nation's Report Card: Science 2011* (NCES 2012–465). Institute of Education Sciences, U.S. Department of Education, Washington, D.C.

National Center for Education Statistics. (2011). *The Nation's Report Card: Trial Urban District Assessment Mathematics 2011* (NCES 2012–452). Institute of Education Sciences, U.S. Department of Education, Washington, D.C.

National Center for Technology Innovation and Center for Implementing Technology in Education (CITEd). (2009). *Technology Supported Science, Technology, Engineering, and Math (STEM) Instruction for Students with Disabilities*. Retrieved from: http://www.ldonline.org/article/35729/

National Council of Teachers of Mathematics (NCTM). (2000). *Principles and standards for school mathematics*. Reston, VA: Author.

National Research Council. (2013). *Next Generation Science Standards: For states, by states*. Washington, DC: The National Academies Press.

New London Group. (1996). A pedagogy of multiliteracies: Designing social futures. *Harvard Educational Review, 66,* 60-92.

O'Connor, C., & Fernandez, S. D. (2006). Race, class, and disproportionality: Reevaluating the relationship between poverty and special education placement. *Educational Researcher, 35*(6), 6-11.

Orosco, M. J., & Klingner, J. K. (2010). One school's implementation of RTI with English language learners: Referring into RTI. *Journal of Learning Disabilities, 43*(3), 269–288.

Pyrch, T., & Castillo, M. T. (2001). The sights and sounds of indigenous knowledge. *Handbook of action research, participative inquiry and practice*, 379-385.

Roth, W. M., & Lee, S. (2002). Scientific literacy as collective praxis. *Public Understanding of Science, 11*, 33–56.

Sandia Mountain Natural History Center (n.d.). Retrieved from: https://sites.google.com/a/aps.edu/smnhc/home

Saunders, W. M., & Goldenberg, C. (1999). Effects of instructional conversations and literature logs on limited-and fluent-English-proficient students' story comprehension and thematic understanding. *The Elementary School Journal*, 277-301.

Schoenfeld, A. (1985). *Mathematical problem solving*. Orlando, FL: Academic Press, Inc.

Schumm, J.C., Vaughn, S., & Leavell, A.G. (1994). Planning pyramid: A framework for planning for diverse students' needs during content instruction. *The Reading Teacher, 47*, 608-615.

Skiba, R. J., Simmons, A., Ritter, S., Kohler, K., Henderson, M., & Wu, T. (2006). The context of minority disproportionality: Practitioner perspectives on special education referral. *Teachers College Record, 108*, 1424–1459.

Snow, C. (2008). What is the vocabulary of science? In A. Rosebery and B. Warren (Eds.) *Teaching science to English language learners* (pp. 71–84). Arlington, VA: NSTA Press.

Sotomayor, K. (2013). Teaching STEM to English Language Learners. *Principal, Jan/Feb*, 40-41.

Thomasian, J. (2011). Building a Science, Technology, Engineering and Mathematics Education Agenda: An update of state actions. NGA Center for Best Practices. Washington, D.C.

Tomlinson, C. (2001). *How to differentiate instruction in mixed-ability classrooms.* Alexandria, VA: Association for Supervision and Curriculum Development. ED 386 301

Torres-Velásquez, D., & Lobo, G. (2004/2005). Culturally responsive mathematics instruction. *Teaching Children Mathematics, 11*(5), 249–255.

U.S. Department of Education (2011). *Office of Special Education and Rehabilitative Services, Office of Special Education Programs, 30th Annual Report to Congress on the Implementation of the Individuals with Disabilities Education Act, 2008.* Washington, D.C.: Author.

U.S. Department of the Interior. (2013). STEM education and employment pathways strategic plan fiscal years 2013—2018. Washington, D. C.

Van de Walle, J. A., Karp. K. S., & Bay-Williams, J. M. (2013). *Elementary and middle school mathematics: Teaching developmentally* (8th ed). Boston, MA: Prentice Education Press.

Van Garderen, D. (2012) Developing representational ability in mathematics for students with learning disabilities: A content analysis of grades 6 and 7 textbooks. *Learning Disability Quarterly, 35*(1): 24-38.

Van Garderen, D., & Montague, M. (2003). Visualspatial representations and mathematical problem solving. *Learning Disabilities & Research, 18*, 246-254.

Warren, B., & Rosebery, A. (2008). Using everyday experience to teach science. In A. Rosebery and B. Warren (Eds.), *Teaching science to English language learners* (pp. 39–50). Arlington, VA: NSTA Press.

Watson, S., & Houtz, L.E. (2002). Teaching science: Meeting the academic needs of culturally and linguistically diverse students. *Intervention in School and Clinic, 37*, 267-268.

Westby, C. (2010). Multiliteracies: The changing world of communication. *Topics in Language Disorders, 30*(1), 64-71.

Westby, C., & Inglebret, E. (2012). Native American and worldwide Indigenous cultures. In D. E. Battle (Ed.), Communication disorders in multicultural and international populations (pp. 76-101). St. Louis, MO: Elsevier.

World Health Organization. (2001). *International classification of functioning, disability, and health (ICF).* Geneva: Author.

In: S.T.E.M. Education
Editor: Satasha L. Green
ISBN: 978-1-63117-806-1

Chapter 12

WORKING WITH CULTURALLY DIVERSE LEARNERS WITH SPECIAL NEEDS IN STEM

Sandra Cooley-Nichols and Adriane Sheffield
University of Alabama, Tuscaloosa, AL, US

ABSTRACT

Several trends emerge from annual educational reports related to culturally and linguistically diverse (CLD) students and school settings. CLD refers to students from racial/ethnic minority groups and students from language minority groups. Many national educational initiatives currently involve science, technology, engineering and mathematics fields (STEM). While STEM is not typically discussed as an approach to address the needs of CLD students with special needs, they are an ideal group to target for STEM instruction. This chapter will discuss the learning styles of CLD students, the importance of STEM instruction for CLD students with special needs and instructional strategies that support STEM instruction for CLD students with special needs.

INTRODUCTION

In 2012, the US Department of Education published the *Condition of Education*. This annual report uses data collected from the US Census Bureau and provides demographic data on school-aged students from Prekindergarten through postsecondary education. There are several trends that emerge from this annual report related to culturally and linguistically diverse (CLD) students and school settings. First, the number of students in the United States from racially, ethnically and linguistically diverse backgrounds continues to increase. Second, this trend towards increased diversity will continue.

Third, while the student body becomes increasingly diverse, the teaching population remains primarily unchanged. These trends combined create an issue within the educational system that must be addressed in order to ensure the success of all students, particularly those who have been identified for special education services.

When looking at the prevalence of CLD students, it is important to define what is meant by CLD. For the purpose of this chapter, CLD refers to students from racial/ethnic minority groups and students from language minority groups. Language minority groups include those who speak a native language other than English. Among CLD students, the most common subgroups include: Black, Hispanic, Asian and English Language Learners [ELL] (Sullivan, 2011). The majority of ELL students in the United States are Hispanic and their first language is Spanish (Hopstock and Stephenson, 2003). According to US Census data, CLD students are becoming the majority of public school populations nationally. While some urban districts have a large concentration of CLD students, even rural schools are dealing with increased diversity (Conroy, 2012). While the student body becomes increasingly diverse, the teaching population remains primarily white and female. As a result, there is frequently a mismatch between the culture of students and the culture of schools. It is because of this mismatch that is important to discuss key factors about each group of students and how they are affected by current special education services.

Blacks

As a group, Black students have experienced the least amount of school success across general, special, and gifted education (Ford, 2012). As a result, they are one of the most studied groups. Blacks currently make up about 15% of the school population (Aud et al., 2010). Yet they are one of the CLD groups that receive the largest number of referrals to special education. In some areas of special education, Blacks are over-represented while in other areas they are under-represented. Over-representation is defined as the occurrence of a greater percentage of minority students in special education programs than in the general school population. According to the Office of Civil Rights (OCR) (2006), while Black students represented 17.13% of the school population, more than 30% were identified as having an intellectual disability; more than 25% were identified with emotional disturbance, 20% with a specific learning disability and 11% with a developmental delay. Black males showed the greatest over-representation in each category. In the area of gifted education, Black students are under-represented by 48% (Ford, 2010).

Hispanic Americans

When looking at Hispanic students in special education, it is important to remember that this classification also includes those who consider themselves Latino and Puerto Rican. Nationally, Hispanic Americans make up approximately 20% of the public school population. However they make up only 15% of those with intellectual disabilities, 11% of those with emotional disturbance, about 21% of those with specific learning disability, and 11% of those with developmental delays. Although national data does not support over-representation in special education, there is a great deal of variation across states (Guiberson, 2012; Sullivan, 2011). There is also variation by disability category. These variations have been attributed by some researchers (Chu, 2011; Irvine, 2012; Sullivan, 2011) to school district size, per capita student spending and increased diversity among student population. As it pertains to gifted education, Hispanic students are underrepresented by 38% nationally (Ford, 2010).

Asian-Americans

Asian-Americans and Pacific Islanders are a combined group in the Census data. Based on the 2006 OCR report, they comprise 4.81% of the student population. As it pertains to disability categories, 2.8% are classified as having an intellectual disability, 1% as having emotional disturbance, less than 2% as having a specific learning disability and a little over 3% as having a developmental delay. When looking at gifted education, 13% of Asian-American/Pacific Islander students are classified as gifted and talented nationally.

English-Language Learners

English Language learners are primarily comprised of Hispanic and Asian-American/ Pacific Islander students. This category also includes students who speak English with difficulty. According to statistics (Aud et al., 2010), approximately 5% of public school students speak a language other than English at home or have difficulty speaking English. There is a large amount of variation in the number of English Language Learners who are identified as needing special education services. The national average is approximately 9% (Sullivan, 2011). ELL students are some of the most difficult to properly identify because of the overlapping issues of under-referral and over-diagnosis related to misunderstanding of their educational needs (Case and Taylor, 2005), weak psycho-educational assessment practices (Figueroa and Newsome, 2006) and ill-defined language assessments (MacSwan and Rolstand, 2006). Students identified as ELL tend to perform poorly on high language academic tasks. This is similar to students identified as LD. This makes it difficult for educators to determine if academic problems for ELL students are language related or cognitive. As a result ELL students risk receiving inappropriate services. It is important to find ways to appropriately address the needs of ELL students across all subject areas, but particularly those with high vocabulary content and more abstract concepts make it critical to make STEM instruction more explicit (Meyer and Crawford, 2011).

In summary, there are numerous issues of overrepresentation and underrepresentation as it pertains to culturally and linguistically diverse students. Overrepresentation is found among Black and ELL students. While overrepresentation does not appear to be an issue for Hispanic students, this varies by region and school district.

When looking at specific categories of disability, Black males are over-represented in every category. As it pertains to underrepresentation, Blacks and Hispanics are significantly underrepresented in gifted education.

Importance of Stem Instruction for CLD Students with Special Needs

Culturally and linguistically diverse students are at greater risk for placement in special education due to a variety of contextual issues (Klingner et al., 2005). These factors include: (a) decision-making processes that determine special education eligibility; (b) special education program placements with varying levels of restrictiveness; (c) administrative decisions that result in disparities; (d) lack of alternative program availability (early interventions, bilingual education, etc.); (e) subtle bias in the referral process; (f) interactions between school location, disability, ethnicity, and density of CLD populations; (g) uneven quality of instruction in general classrooms; and (h) varying discipline policies (i.e. suspensions). When looking at all of the additional factors which affect CLD students, it is important to identify culturally responsive teaching practices that support their learning, particularly when they have special needs.

To effectively address the needs of CLD students with disabilities, it is important to look at culturally responsive teaching practices, policies and initiatives. Cultural responsiveness is based on the belief that "culturally and linguistically diverse students can excel in academic endeavors when their culture, language, heritage and experiences are valued and used to facilitate and support their learning and development *and* they are provided access to high quality programs, teachers and resources" (Gay, 2000).

Many national educational initiatives currently involve the science, technology, engineering and mathematics fields (STEM). This national focus on STEM fields is not only an educational issue, but an economic one. Many current professions require a certain level of STEM understanding (Basham and Marino, 2013).

Without the knowledge base offered by STEM fields, students of all backgrounds are at a disadvantage and CLD students with disabilities are at an even greater disadvantage. When thinking about the STEM fields, it is important to remember that STEM moves beyond delineated subjects. STEM is an interdisciplinary approach to learning that emphasizes problem-solving and critical thinking through real-world applications (Basham, Israel and Maynard, 2010; Israel, Maynard and Williamson, 2013).

As a progressive nation that is focused on technology, innovation and economic growth, it is imperative that all students have the capacity to think critically and flexibly while solving problems in innovative ways. While STEM is not typically discussed as an approach to address the needs of CLD students with special needs, they are an ideal group to target for STEM instruction. There are certain learning characteristics for CLD students that naturally align with STEM instruction. Boykin (1994; Boykin, Tyler and Miller, 2005) identified nine learning characteristics that are culturally relevant for African-American students. These include: (1) spirituality, (2) harmony, (3) affect, (4) communalism, (5) movement, (6) verve, (7) expressive individualism, (8) orality, and (9) social time orientation (Ford, 2012).

Spirituality is a belief in a higher power. Harmony is acknowledgement of a connection between people and the environment. Affect is sensitivity to emotional expressiveness. Communalism is an orientation to family or community. Movement is a tendency toward active learning experiences. Verve refers to a high level of sensitivity to physical or sensory stimulation. This can include intensity, variability and density of sensory stimulation.

Expressive individualism is an appreciation of one's unique style of expressiveness. Orality is a preference for verbal forms of communication as well as verbal bluntness. Social time orientation is a focus on events rather than the passage of time.

Similar to Black students' sense of communalism, 'familisimo' is an emphasis on the needs of the family over individual needs. Hispanic students tend to be more cooperative and social and prefer working collaboratively. Of these learning characteristics, several appear to be conducive for STEM instruction.

Harmony is a natural recognition of one's connection with the environment. Basham and Marino (2013) discuss STEM as being about engineering design, which is primarily about problem solving with consideration of constraints. By constraints, engineers look at issues that may impact possible solutions. This could include budgetary, political, and time constraints. By allowing CLD students with disabilities to participate in engaging learning activities that address problems within their own communities, one can acknowledge and incorporate student's connection with the environment or sense of harmony.

For many, this is also away to address the spiritual connection that is common with many CLD students without creating religious conflicts.

Communalism or a preference for working collaboratively and interdependently is also a common learning characteristic of CLD students with and without special needs. STEM is conducive to large amounts of group interaction because it emphasizes student-centered collaborative learning processes (Israel, Maynard, and Williamson, 2013; Basham and Marino, 2013). This affords CLD students with special needs the opportunity to learn about their world in a more authentic and meaningful way. It does however require educators to support student learning through specific vocabulary instruction and scaffolding of abstract concepts.

In addition to harmony and communalism, movement and verve are common learning characteristics of CLD students (Ford, 2012; Boykin, 1994; Boykin, Tyler and Miller, 2005). Movement and verve involve an orientation to tactile and kinesthetic activities. CLD students tend to be kinesthetic learners, some with high levels of energy. STEM tasks tend to engage students in more kinesthetic activities such as creation and construction oriented tasks. This gives all CLD students including those with special needs an appropriate classroom space to exercise their need for movement.

The fact that STEM instruction creates an ideal backdrop for CLD students' learning characteristics exemplifies the importance of this discussion at this point in time. As federal initiatives encourage, support and fund STEM opportunities across grade levels and institutions, it is imperative that we identify ways to support CLD students effectively for STEM learning and professions.

Problem- solving is the rudimentary foundation of education in the STEM areas. As previously mentioned in this chapter, CLD students are most successful when learning is associated with "community". Community in this instance involves allowing the students to work together to solve problems and supporting the development of personal identification with instructional content and skills. When applied to STEM knowledge and skills, this personal identification gives students a familiar context to draw upon when new concepts, theories, and applications are introduced. Additionally when students are able to build a STEM focused personal identity they are able to garner a deeper understanding of the material through concrete examples (Burke, 2007; Brayboy and Castagno, 2008). This connection is a necessary phase of the learning process that must be facilitated for students with learning and intellectual disabilities. Students with learning or intellectual disabilities experience varying types and degrees of difficulty in reading, mathematics, receptive and/or expressive communication, organization, concentration, and generalization. For years an assumption has been made that these differences nullify students with learning or intellectual disabilities abilities to greatly benefit from STEM related instruction. Consequently when disabilities were paired with cultural and linguistic differences, STEM education was considered to be beyond the reach of CLD students with disabilities.

Fortunately, there are a variety of pedagogical approaches that support both the identified learning characteristics of CLD students with disabilities as well as the instructional delivery of STEM related concepts and knowledge. Several of the most widely used instructional strategies that are used to teach STEM related concepts and knowledge is explanation, argumentation, nature of science, and explicit instruction (Stage, Asturias, Cheuk, Daro, and Hampton, 2013; Meyer and Crawford, 2011). Each of these instructional strategies is a core idea within the National Research Council [NRC] (2011) report, *A Framework for K-12 Science Education: Practices, Crosscutting Concepts, and Core Ideas*. Reiser, Berland, and Kenyou (2012) described the inquiry based scientific practices, which are included in the framework as: (a) questioning and defining problems, (b) developing and implementing models, (c) planning and conducting investigations, (d) analyzing and interpreting data, (e) using mathematical and computational thinking, (f), creating explanations and designing solutions, (g) engaging in fact based arguments, and (h) gathering, authenticating, and disseminating information. *The Next Generation Science Standards* (Achieve, 2013) that were recently published are closely aligned with the framework and include these same instructional strategies as well as other information. Even more, the science Framework overlaps with the other STEM areas, technology, engineering, and mathematics. These areas are integrated into the science framework and reflected in the following pedagogies.

Pedagogy

Explanation, argumentation, nature of science and explicit instruction are pedagogical approaches, which have been designed to increase student engagement and mastery of STEM related concepts and skills (Meyer and Crawford, 2011; Braaten and Windschitl, 2011; Reiser, Berland, and Kenyon, 2012). The framework (NRC, 2011) describes explanation as "accounts that link scientific theory with scientific observations or phenomena."

Explanation is a common skill that students are expected to practice in school settings. When students present differing perspectives or opposing positions, they are often given the opportunity to "explain" themselves. While there is variance in implementation of the practice of *explaining,* in general, teachers typically view *explain* as providing clarification, stating why something happened, or giving an account/justification of one's beliefs (Braaten and Windschitl, 2011). More important are the specific pedagogical skills that teachers must possess in order to facilitate student delivery of defensible explanations. First, teachers must have a clear understanding of explanation that manifests in the ability to link a chain of rational concepts with the phenomenon to be explained (Reiser, Berland, and Kenyon, 2012). Secondly, the teacher must be familiar with concrete scientific examples that occur in the lives of his/her students and be able to match these occurrences with the theory that is being taught. The National Research Council (2011) refers to *explanation* as linking scientific theory with scientific observation or phenomena and presenting a statement that identifies the process of cause and effect.

Therefore, teachers must fully understand the practice of *explanation* and the pedagogy of teaching culturally and linguistically diverse students (Gay, 2010; NRC, 2011; Watson and Houtz, 2002). Teachers must be able to identify and facilitate their students' identification of scientific phenomena. Teachers must be able to logically sequence a chain of events that are connected to the identified phenomena and teach their students to implement this process. Likewise, teachers must have evidence based knowledge of learning styles and practices representative of culturally and linguistically diverse populations.

Explanation gives students the opportunity to present topics and engage in meaningful discourse that leads to refinement and expansion of the ideas, which they used to support the original phenomena (NRC, 2011). It also allows them to create and invent solutions that address needs within their immediate surroundings. However, students must be guided by a teacher that has mastered the instructional and the facilitative process of explanation.

Argumentation is an essential component of the social construction of scientific knowledge. It provides an avenue for students to build upon their preexisting scientific knowledge and to evaluate the criteria used to establish scientific theories (Reiser, Berland, and Kenyon, 2012). Argumentation must be included in the process of acquiring and comprehending scientific knowledge in order to improve the level of public science literacy (NRC, 2011).

The classroom is the students' and educators' laboratory; it gives a space to construct arguments. Teachers need the following pedagogical skills to organize argumentative discourse within the classroom. Teachers must first have classroom rules that structure the communication and support expression of thought with dignity, respect, and equity of voice. The National Research Council (2011) refers to arguing as scientists utilizing research based knowledge to make a reasonable justification about an issue or phenomena. Other scientists then attempt to contest the claim by identifying its weakness or limitations.

Thus, teachers must be able to facilitate students' construction of scientific arguments that support their claims (NRC 2011; Reiser, Berland, and Kenyon, 2012). Teachers must be able to identify and teach the students to identify possible limitations in scientific arguments, which are developmentally and academically appropriate, and use logical and research based evidence in discussion. Teachers must be able to identify and teach the students to identify errors in their own arguments and adjust and improve them based on the feedback.

Argumentation improves students' ability to think critically. They are required to make a hypothesis, find research which supports the foundation of the hypothesis, provide enough related knowledge that it makes the idea believable, and engage in spontaneous discussion that is constructed from knowledge that will build consensus. In order for students to be able to perform at this level, teachers of science must understand and be able to facilitate the practice of argumentation.

Nature of Science (NOS) is a term commonly used in science education that addresses a holistic approach to science. NOS includes methods and social-ethical dimensions of science [inclusive of CLD consideration], and the intersection of science and society (Develaki, 2012; Grimberg and Gummer, 2013). While NOS conceptually supports the instruction of CLD students, curricular content and implementation must be delivered, conducted intentionally. There are specific pedagogical knowledges and skills that teachers must possess in order to successfully provide NOS grounded instruction. Teachers must have a content base in the area of instruction, which includes understanding of scientific principles and concepts as well as representative culturally relevant accounts that occur in their students' lives. Teachers must have a clear understanding of culturally relevant accounts that are void of stereotypes and assumptions.

NOS allows all students to participate in science education from an angle which allows them to use their background knowledge and experience to develop a better understanding of the information that they are learning. This unique yet logical perspective makes science learning meaningful and demonstrates how science affects the students' daily lives. Perhaps more importantly, students learn how their actions positively and negatively impact society.

Hopefully, this approach will encourage the students to identify ways to improve the world in which we live.

When these pedagogical approaches and others are used interchangeably, CLD students with disabilities are supported to master STEM education at their highest level of learning potential (Villanueva and Hand, 2011; Spaulding and Flannagan, 2012; Miller, 2012).

It is notable to iterate that selecting the appropriate pedagogical approach correlates with the content and skills that are being taught.

The figure below depicts the continual overlap of practices infused in the K-12 science education framework that was designed by the National Research Council (2011).

While the framework was designed for "science", the STEM areas have been considered the hard sciences for many years. Therefore it is a fair assumption that the overlap shown in this illustration is intentional.

K – 12 Teacher Strategies and Resources

STEM education is an active process (National Research Council, 1996). Students are actively engaged in researching established theories to obtain understanding, to discover new paths of inquiry, to generalize findings to daily practice, and to improve the overall wellbeing of society (NCR, 2011). This type of instruction is based upon hands-on learning. CLD students with disabilities thrive in active, hands-on learning environments, especially those that draw upon their backgrounds and experiences (Spaulding and Flannagan, 2012; Miller, 2012; Villanueva and Hand, 2011). This connection provides CLD students with disabilities access to relevant examples that assist them in understanding the associated vocabulary, text, and oral/digital/environmental information. Thus when these considerations are made and properly, addressed, every science classroom has the potential to become a successful learning environment for CLD students with disabilities.

The following information provides a snapshot of the teaching strategies that are employed in the STEM subject areas. Although there is considerable overlap, science and mathematics are presented independently, while technology and engineering function as threads that are integrated.

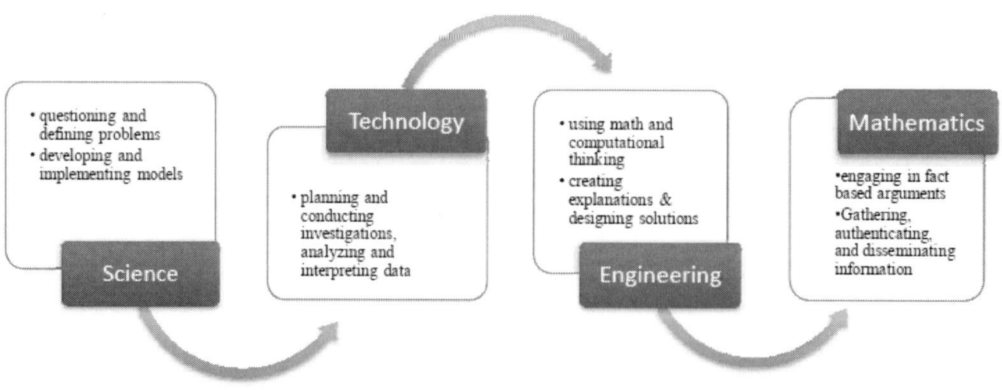

Figure 1. The overlap of practices infused in the K-12 science education framework.

It is important to note that the strategies and techniques discussed are not specific to one content area. They are interchangeable. The decision of use should be made with the student, the content, the setting, and the learning expectation in mind.

Science. Interactive classroom demonstrations require students to be actively engaged (Gross, 2002). Teachers and/or students may conduct the demonstration. Typically the teacher facilitates the demonstration while allowing the students to actively participate (Reeder, 2010). The teacher asks opened questions in order to elicit student response. The teacher provides probes and cues, so that the students will give more detailed explanations of their commitment to a specific chain of reasoning. The teacher facilitates the students' learning of how to determine whether a source is reputable and to draw conclusions from the evidence gathered. Interactive classroom demonstration aligns with CLD pedagogy.

Students are given an outlet for hands-on learning, which incorporates examples that occur in their daily activities (Gay, 2010).

Mathematics. Graphic Organizers may be used to assist students with organization and memorization of content (Horton, Lovitt, and Bergerus, 1990). Culturally and linguistically diverse students with disabilities may benefit greatly from the use of graphic organizers, especially when learning skills such as how to use diagrammatic forms to represent problem situations (Jitendra, 2002). Graphic organizers are helpful in identifying the necessary operation(s) that must be implemented in order to find a solution to a problem (Banks, 2010).

Additionally, the graphic organizers give CLD students with disabilities an outlet for expression. The students are given a way to explain the information that they have learned from a more creative canvas. The graphic organizers may also be utilized in group activities.

This structure supports CLD pedagogy by incorporating a community-based assignment (Gay, 2010). Furthermore, giving a graphic organizer based activity to a trained peer tutor that is paired with a CLD student with a disability addresses the CLD student with a disability's need for extra support. It also gives the typically developing peer tutor an opportunity to engage in explanation, which enhances the scientific learning process.

Another option that may be implemented in the process of learning mathematics is using manipulatives. Manipulatives are not restricted to use in elementary settings; they are available in a variety of groupings which are designed for differing levels of complexity (Miller et al., 2011). Research indicates that CLD students with disabilities have achieved positive outcomes when they are allowed to use manipulatives to learn mathematics.

In order for the manipulatives to enhance students' mathematics learning, the teacher must fully understand the concepts behind utilizing manipulatives and the implementation process (Powell, 2010; Saunders, et al., 2013).

Engineering and Technology. Are themes that are interwoven throughout STEM education (NRC, 2011). The conceptual framework of STEM education is mainly based on engineering theories. Students are expected to ask questions, explore alternatives, design/plan solutions, and create/build options (NRC, 2011). This basic but powerful tenant can be found in science, technology, and mathematics instruction and practice (Miller, 2011, Villanueva and Hand, 2011, Reiser, Berland, and Kenyon, 2012).

In general engineering is comprised of science, technology, and mathematics content, it is the expansion and physical product creation that makes this area uniquely different. Thus, the pedagogy and teaching strategies and resources that are presented in this chapter are relevant to engineering teachers. Possibly most important is the fact that engineering is linked to the learning strengths of CLD students (Ogbu, 1992; Gay, 2010).

Characteristics of CLD students include hands-on learners [i.e., create/build], creative [i.e., plan/design], and community oriented [i.e., ask questions and explore alternatives].

Technology is a foundational component of society as well as STEM education. Although many teachers and students do not get the most out of technology for academic learning, the majority of these individuals are deeply submerged in technology usage for personal/social interaction. Teachers must be able to bridge the gap between casual technology usage and the use of technology for formal learning, specifically expand beyond tutorial or practical application (Basham and Marino, 2013; Israel, Maynard, and Willamson, 2013).

Technology is a living medium. It allows students to collaborate with others in differing schools, cities, states, countries. It gives students and teachers resources for enhancement and tutorial purposes [e.g., electronic books, digital note takers, content related software programs, search engines] (Carnahan et al., 2012); but it also provides an avenue for research, creation, inquiry, and exploration of established principles. Technology supports student involvement and can be used to individualize their education based on interests, culture, background, and experiences (Grimberg and Gummer, 2012). Teachers may use technology to deliver information and prescribe established programs for their students [i.e., students are consumers] and teachers may use technology to facilitate student discovery and invention [i.e., students are producers of knowledge].

Differentiating STEM Instruction

Print rich environments. Are essential to the academic vocabulary development of CLD students with disabilities (Cummins, 2009). Although print rich environments are typically associated with early childhood and elementary education, it is also imperative that CLD students with disabilities on middle and secondary levels have frequent meaningful interaction with academic vocabulary (Gay, 2010). Academic vocabulary is infrequently used in familial conversation and casual communication among the social groups that CLD students with disabilities normally interact. CLD students with disabilities not only need to see and read the printed vocabulary but they must engage in report writing, research, and discussion (i.e., explanation and argumentation), which include academic vocabulary (Spaulding and Flannagan, 2011; Villanueva and Hand, 2011).

Identity affirmation. Is strongly linked to literacy engagement (Cummins, 2009). It is important to emphasize the link that identity affirmation has to content literacy. It is crucial to the academic success of CLD students with disabilities that teachers systematically make concrete connections between the knowledge and skills that are being introduced during instruction and the activities that these students participate in on a daily basis. According to Ogbu (1992) CLD students with disabilities have a larger investment in the learning process when teachers acknowledge their talents, display respect for their culture and background, and make learning relevant.

Activate prior knowledge. Supports higher levels of comprehension in STEM areas. As previously mentioned, it is necessary to promote understanding of STEM content on a personal basis. The phenomena that are presented in an academic format must be matched with an example that occurs in the day-to-day experiences of the CLD student (Cummins,

2009). Even more we must remember that as each student with a disability needs are considered individually, each CLD student is unique.

There are some similarities; however, in order to have the greatest effect on student learning teachers have to get to know and understand the culture, background, experiences, and learning styles of each individual student (Ogbu, 1992; Gay, 2010). This knowledge allows the teacher to connect with the student, which directly reflects the *community* system that has been documented in CLD research.

When the connection [i.e., students know that you care] is made with CLD students, teachers are capable of eliciting high levels of engagement, active participation, critical thinking, analysis, and expansion of thought… positive outcomes (Watson and Houtz, 2002).

Scaffolding. A technique, which provides continual success for CLD students with disabilities as the instructional information and material gradually becomes more challenging (Watt, Therrien, Kaldenberg, and Taylor, 2013; Spaulding and Flannagan, 2012; Israel Maynard and Williamson, 2013). Knowledge and skills are presented with increasing complexity in a step-by-step or phase-by-phase format. Often times in special education, we rely upon the "I do, we do, you do" method. While we know that this method is effective, it is important that in addition to employing strategies such as direct instruction (Mason and Hedin, 2011; Smith, Spooner, Jimenez, and Browder, 2013) and reciprocal teaching (King and Johnson, 1999; Mason and Hedin, 2011) students are given the opportunities to explore and create. Active exploration and engagement in creativity, encourages critical thinking and supports STEM learning. Scaffolding paired with the aforementioned strategies gives CLD students with disabilities the ability to learn STEM knowledge and skills in a safe environment, which addresses their individual learning needs and attends to their cultural identities.

Differentiating STEM instruction. Is imperative to the success of CLD students with disabilities because it makes the curriculum accessible (Ladson-Billings, 1995). Students have different learning styles, exceptionalities, cultures, languages, social norms, socioeconomic statuses, etc. It is important that both the student and the teacher know the student's learning style(s) [verbal-linguistic, visual-spatial, logical mathematical, kinesthetic, musical, naturalist, interpersonal, and intrapersonal] (Gardner, 2010). When teachers know their students, they are able to differentiate content, process, and product according to their students' readiness, interest, and learning styles (Tomlinson, 1999).

REFERENCES

Achieve, Inc. (2013). *Next generation science standards: For states, by states*. Retrieved from: http://www.nextgenscience.org/

Argumentation. (n.d.). *Dictionary.com Unabridged*. Retrieved from: http://dictionary.reference.com/browse/argumentation.

Aud, S., Hussar, W., Johnson, F., Kena, G., Roth, E., Manning, E., Wang, X., and Zhang, J. (2012). *The Condition of Education* 2012 (NCES 2012-045). US Department of Education, National Center for Education Statistics. Washington, DC. Retrieved from http://nces.ed.gov/pubsearch.

Aud, S., Fox, M. A. and Kewal Ramani, A. (2010). *Status and trends in the education of racial and ethnic groups.* (NCES 2010-015) Washington, DC: US Department of

Education, National Center for Education Statistics. Retrieved from http://nces.ed.gov/pubs2010/2010015.pdf.

Banks, T. (2010). *Using graphic organizers in mathematics.* A presentation for the STRIVE II: Professional Development (Fayette County Schools), at the University of Memphis Fogelman Executive Center, Memphis, Tennessee.

Basham, J. D. and Marino, M. T. (2013). Understanding STEM education and supporting students through universal design for learning. *TEACHING Exceptional Children,* 45 (4), 8-15.

Braaten, M. and Windschitl. (2011). Working towards a stronger conceptualization of scientific explanation for science education. *Science Education,* 95(4), 639-669.

Brayboy, B. M. B. and Castagno, A. E. (2008). How might native science inform "informal science learning?" *Cultural Studies of Science Education,* 3(3), 731-750.

Burke, C. (2007). Neighborhood science stories: Bridging science standards and urban students' lives. *Teaching Education,* 18(4), 357-367.

Carnahan, C. R., Williamson, P. S., Hollingshead, A., and Israel, M. (2012). Using technology to support balanced literacy for students with significant disabilities. *TEACHING Exceptional Children,* 45(1), 20-29.

Case, R. E. and Taylor, S. S. (2005). Language difference or learning disability? Answers from a linguistic perspective. *The Clearinghouse*, 78, 127-130.

Chu, S. (2011). Perspectives in understanding the schooling and achievement of students from culturally and linguistically diverse backgrounds. *Journal of Instructional Psychology,* 38 (3), 2019-209.

Conroy, P. W. (2012). Collaborating with cultural and linguistically diverse families of students in rural schools who receive special education services. *Rural Special Education Quarterly,* 31(3), 24-27.

Cummins, J. (2009). Transformative multiliteracies pedagogy: School-based strategies for closing the achievement gap. *Multiple Voices for Ethnically Diverse Exceptional Learners,* 11(2), 38-56.

Develaki, M. (2012). Integrating scientific methods and knowledge into the teaching of Newton's Theory of Gravitation: An instructional sequence for teachers' and students' nature of science education. *Science and Education,* 21, 853-879.

Figueroa, R. A. and Newsome, P. (2006). The diagnosis of LD in English learners: Is it nondiscriminatory? *Journal of Learning Disabilities,* 39, 206-214.

Ford, D. Y. (2010). Underrepresentation of culturally different students in gifted education: reflections about current problems and recommendations for the future. *Gifted Child Today,* 33(3), 31-35.

Ford, D. Y. (2012). Culturally different students in special education: looking backward to move forward. *Exceptional Children,* 78(4), 391-405.

Gardner, H. (1999). *Intelligence reframed: Multiple intelligences for the 21st century.* New York, NY: Basic Books.

Gay, G. (2010). *Culturally responsive teaching: Theory, research and practice.* (2nd ed.). New York: NY: Teachers College Press.

Grimberg, B. I. and Gummer, E. (2013). Teaching science from cultural points of intersection. *Journal of Research in Science Teaching,* 50(1), 12-32.

Gross, J. (2002). Seeing is believing: Classroom demonstrations as scientific inquiry. *Journal of Physics Teachers Education Online.* Retrieved from: http://www.phy.ilstu.edu/pte/311 content/demos/demos_as_inquiry.pdf.

Guiberson, M. (2009) Hispanic representation in special education: patterns and implications. *Preventing School Failure,* 53(3), 167-176.

Hopstock, P. J. and Stephenson, T. G. (2003). *Descriptive study of services to LEP students and LEP students with disabilities.* National Clearinghouse for English Language Acquisition.

Horton, S. V., Lovitt, T. C. and Bergerud, D. (1990). The effectiveness of graphic organizers for three classifications of secondary students in content area classes. *Journal of learning disabilities,* 23(1), 12-22.

Irvine, J. J. (2012). Complex relationships between multicultural education and special education: An African American perspective. *Journal of Teacher Education,* 63(4), 268-274.

Israel, M., Maynard, K. and Williamson, P. (2013). Promoting literacy-embedded, authentic STEM instruction for students with disabilities and other struggling learners. *Teaching Exceptional Children,* 45(4), 18-25.

Jitendra, A. K. (2002). Advanced story map instruction: Effects on the reading comprehension of students with learning disabilities. *The Journal of Special Education,* 331(1), 2-17.

Klingner, J. K., Artiles, A. J., Kozleski, E., Harry, B., Zion, S., Tate, W., Durian, G. Z., and Riley, D. (2005). Addressing the disproportionate representation of culturally and linguistically diverse students in special education through culturally responsive educational systems. *Education Policy Analysis Archives,* 13(38). Retrieved from http://epaa.asu.edu/epaa/v13n38/.

Ladson-Billings, G. (1995).Toward a theory of culturally relevant pedagogy. *American Educational Research Journal,* 32, 465-491.

MacSwan, J. and Rolstad, K. (2006). How language proficiency tests mislead us about ability: Implications for English language learner placement in special education. *Teachers College Record,* 108, 2304-2328.

Mason, L. H. and Hedin, L. R. (2011). Reading science texts: Challenges for students with learning disabilities and considerations for teachers. *Learning Disabilities Research and Practice,* 26(4), 214-222.

Meyer, X. and Crawford, B. A. (2011). Teaching science as a cultural way of knowing: merging authentic inquiry, nature of science, and multicultural strategies. *Cultural Studies of Science Education,* 6, 525-547.

Miller, B. (2012). Ensuring meaningful access to the science curriculum for students with significant cognitive disabilities. *Teaching Exceptional Children,* 44(6), 16-25.

Miller, S., Stingfellow, J. L., Kaffar, B. J., Ferriera, D., and Mancl, D. B. (2011). Developing computation competence among students who struggle with mathematics. *TEACHING Exceptional Children,* 44(2), 38-46.

National Research Council (NRC). (1996). *National science education standards.* Washington, DC: National Academy Press.

National Research Council (NRC). (2011). *A Framework for K-12 Science Education: Practices, crosscutting concepts, and core ideas.* Washington, DC: National Academy Press.

Ogbu, J. U. (1992). Understanding cultural diversity and learning. *Educational Researcher,* 21, 5-14 and 24.

Powell, A. (2010). *Fractions II, two, too, to???* A presentation for the STRIVE II: Professional Development (Fayette County Schools), at the University of Memphis Fogelman Executive Center, Memphis, Tennessee.

Reeder, A. (2010). *The WOW factor: Classroom development as scientific inquiry.* A presentation for the STRIVE II: Professional Development (Fayette County Schools), at the University of Memphis Fogelman Executive Center, Memphis, Tennessee.

Reiser, B. J., Berland, L. K. and Kenyon, L. (2012). Engaging students in the scientific practices of explanation and argumentation: Understanding a framework for k-12 science education. *The Science Teacher,* 34-39.

Saunders, A. F., Bethune, K. S., Spooner, F., and Browder, D. (2013). Solving the common core equation: Teaching mathematics CCSS to students with moderate and severe disabilities. *Teaching Exceptional Children,* 45(3), 24-33.

Smith, B. R., Spooner, F., Jimenez, B. A., and Browder, D. (2013). Using an early science curriculum to teach science vocabulary and content to students with severe developmental disabilities. *Education and Treatment of Children,* 36(1), 1-31.

Spaulding, L. S. and Flannagan, J. S. (2012). DIS_2ECT: A framework for effective inclusive science instruction. *Teaching Exceptional Children,* 44(6), 6-14.

Stage, E. K., Asturias, H., Cheuk, T., Daro, P. A., and Hampton, S. B. (2013). Opportunities and challenges in next generation standards. *Science,* 340(19), 276-277.

Sullivan, A. L. (2011). Disproportionality in special education identification and placement of English language learners. *Exceptional Children,* 77(3), 317-334.

Tomlinson, C. A. (1999). *The differentiated classroom: Responding to the needs of all learners.* Alexandria, VA: Association for Curriculum and Development.

Villanueva, M. G. and Hand, B. (2011). Science for all: Engaging students with special needs in and about science. *Learning Disabilities Research and Practice,* 24(4), 233-240.

Watson, S. M. R. and Houtz, L. E. (2002). Teaching science: meeting the academic needs of culturally and linguistically diverse students. *Intervention in School and Clinic,* 37(5), 267-278.

Watt, S. J., Therrien, W. J., Kaldenberg, E., and Taylor, J. (2013). Promoting inclusive practices in inquiry-based science classrooms. *TEACHING Exceptional Children,* 45 (4), 40-48.

In: S.T.E.M. Education
Editor: Satasha L. Green
ISBN: 978-1-63117-806-1

Chapter 13

TRANSITIONING AND PREPARING LEARNERS WITH SPECIAL NEEDS INTO STEM CAREERS

Cari Dunn, Karen Rabren, Melody Russell, Cindy Massey and Michele Martin[1]
Auburn University, Alabama, US

ABSTRACT

Careers in the science, technology, engineering, and mathematics (STEM) fields have many benefits, including good salaries, high job satisfaction, and a strong employment outlook. Moreover, in this increasingly technological society it is critical that we prepare all students and promote equity in STEM representation. Unfortunately persons with disabilities are underrepresented in STEM fields relative to persons without disabilities.

There are a number of reasons for the underrepresentation of persons with disabilities in STEM fields and degree programs which include low expectations, lack of role models, and fewer opportunities for access to valuable science learning experiences. The evidence-based practices that have emerged in the transition research for students with disabilities over the past several decades provide a strong framework for the development of individualized and systematic program planning that will encourage and prepare students with disabilities for STEM careers. The purpose of this chapter is to shed light on strategies that promote and enhance the participation of persons with disabilities in STEM degree programs and careers. A brief overview of the outcomes of adults with disabilities and the legal requirements related to the transition for youth with disabilities will provide the context for practices for transitioning learners with disabilities into STEM.

The chapter concludes with the identification of key transition practices that support STEM career planning and preparation. These practices include individualized and active student involvement in planning, participation in general education, teaching strategies and accommodations, self-determination, technology, career awareness and career/technical education, experiential development, mentoring, collaboration, and postsecondary connecting activities.

INTRODUCTION

The workforce of tomorrow will be dominated by science, technology, engineering, and mathematics (Kelso, 2011). These fields are the foundation for our current economy and future economic growth (Grossman, 2008). STEM fields are diverse, including such areas as chemistry, computer sciences, ecology, and agricultural sciences. Graduates of STEM programs can find jobs as chemists, statisticians, computer programmers, mechanical engineers, teachers, and even music data analysts. The outlook for STEM-related jobs, as well as non-STEM jobs that require a STEM-capable foundation, is favorable. For example, it is projected that the rate of growth in STEM jobs will increase by about 18 %, which is close to double the growth rate for non-STEM jobs, and the earnings of STEM workers is about 26% higher than non-STEM workers (Schiavelli, 2011). As well, workers in many STEM fields (e.g., engineering) consistently report high levels of job satisfaction.

Considering the many benefits of a career in STEM, it seems that a career in this area would be a good choice for many students with disabilities. However, individuals with disabilities are underrepresented in STEM fields. For example, 11.4% of undergraduates in STEM programs have disabilities, but only 5% of the STEM workforce has disabilities (CPST, 2008). Reasons for underrepresentation are related to a range of factors, including K-12 educational issues (e.g., low expectations, inadequate academic preparation), postsecondary education-related issues (e.g., high rates of individuals with disabilities leaving college without earning a degree), and employment issues (e.g., inaccessible environments) (Alston & Hampton, 2000; Alston, Bell, Hampton, 2002; Eriksson, Welander & Grund, 2007; Madaus, Foley, McGuire, and, Ruban, 2002; Nagle, Marder, and Schiller, 2009; Price, Gerber, and Mulligan, 2007). One way to increase the number of students with disabilities pursuing and succeeding in STEM is through individualized transition planning during the secondary school years.

TRANSITION PLANNING AND STUDENTS WITH DISABILITIES

Educating all students with disabilities was not mandated until 1975 with the passage of P.L. 94-142, the Education for All Handicapped Children Act; and consideration of how students with disabilities fared as young adults received little attention until the mid1980s. Results of studies examining the outcomes of youth and young adults with disabilities highlighted the difficulties many experienced in the transition from secondary education to adulthood. This resulted in legislation that required a more systematic and individualized approach to helping prepare school-aged youth with disabilities for their futures.

Outcomes of Individuals with Disabilities

The outcomes of adults with disabilities have been studied through follow-up and follow-along studies for over three decades. The results of these studies are important because they provide valuable information about the lives of young adults with disabilities, which in turn plays an important role in influencing public policy and practice (Johnson, 2012). Even though studies reveal the young adult outcomes have improved over time on several variables, in general, these individuals continue to experience poorer post-school outcomes than their peers without disabilities and are not achieving desired transition-related outcomes.

Individuals with disabilities frequently experience difficulties in postsecondary education participation and retention, employment, participation in community and leisure activities, and dependency on parents (Benz, Lindstrom, & Yovanoff, 2000; Markel & Barclay, 2009; Newman, Wagner, Cameto, & Knokey 2009; Newman, et al., 2011; Sanford, Newman, Wagner, Cameto, Knokey, and Shaver, 2011; Scanlon and Mellard, 2002). For example, a study by Sanford, et al. (2011) found that youth with disabilities were less likely than their peers without disabilities to enroll in postsecondary education (55% vs. 62%, $p<.01$) and less likely to complete programs they started (38% vs. 51%, $p<.01$). In the area of employment, the Bureau of Labor Statistics (June 2012) reported that the workforce participation rates for people without disabilities was 77.7 % compared to 32.1% for people with disabilities (U.S. Department of Labor, 2012). In a July 2012 report, Senator Tom Harkin, Chairman of the Committee on Health, Education, Labor and Pensions, emphasized how workers with disabilities have been hit harder by the recession than workers without disabilities and have been slower to recover from losses experienced from the recession. As well, he noted the underemployment of workers with disabilities. Workers with disabilities are far more likely to be employed part-time (32% vs. 19%) and to have lower median annual earnings ($19,500 vs. $29,997) (Disability Statistics and Demographics Rehabilitation Research and Training Center, 2011).

As a result of such outcomes, much attention has been directed toward what can be done during the school years to better prepare students with disabilities for transition from high school into adulthood. Federal legislation has addressed this need by requiring that transition services be included in the Individualized Education Programs (IEPs) of students with disabilities, and research has been conducted to identify those factors that contribute to positive outcomes for young adults with disabilities.

Overview of the Transition Process

People experience transitions throughout their lifetimes. There are many points of passage along this developmental trek, but for many, the transition that occurs between adolescence and young adulthood can be especially challenging. These can be awkward and turbulent years, because individuals are not quite an adult, yet are no longer a child. This time of development is characterized by Halpern (1992) as "a period of *floundering* that occurs for at least the first several years after leaving school as adolescents attempt to assume a *variety* of adult roles in their communities" (p. 203).

The transition from adolescence to young adulthood can be difficult for many young people, but individuals with disabilities and other special needs face additional obstacles and challenges, simply due to the presence of their disability. Unfortunately, persons with disabilities are often marginalized and considered incapable of assuming productive lives, and expectations typically do not include challenging and rewarding careers such as those within the STEM fields. Barriers associated with gaining employment or continuing education can be so great, that specialized instruction and/or additional assistance may be needed. In order to address these barriers and provide appropriate education, training, and services for these youth and young adults, a national movement known as *transition* began. The goal of this movement is to provide young people with disabilities the support and services they need to become, to the greatest extent possible, independent and contributing members of society.

Transition services were first mandated by the Individuals with Disabilities Education Act of 1990. The definition of transition services included in this legislation and later revised in subsequent amendments provides federal directives for the delivery of these services to youth with disabilities (See Table 1 for definition of transition services). According to this legislation, transition services must be delivered as a "coordinated set of activities." According to this definition the provision of transition services is larger in scope than what can be accomplished by one agency or group of professionals. Consequently, in order to deliver transition services successfully, coordination is needed and requires (a) well-defined roles and responsibilities of service providers and other stakeholders; (b) established methods of communication and information sharing; and (c) regularly scheduled times for developing, measuring, and planning the progress and effectiveness of implementing a shared transition service delivery system.

The foundation of transition planning is the goals students with disabilities make for their lives after high school much like their peers without disabilities, such as continuing their education and training, going to work, improving their social lives, and becoming more independent (Cameto, Levine, & Wagner, 2004). In order to help students with disabilities achieve their desired goals, transition programs and services must address students' needs in these areas.

Some students with disabilities may choose to attend a post-secondary education institution, whether it is a technical school, a 2-year community college or 4-year university.

Table 1. Transition Defined

Transition services.--The term `transition services' means a coordinated set of activities for a child with a disability that—

(A) is designed to be within a results-oriented process, that is focused on improving the academic and functional achievement of the child with a disability to facilitate the child's movement from school to post- school activities, including post-secondary education, vocational education, integrated employment (including supported employment), continuing and adult education, adult services, independent living, or community participation;
(B) is based on the individual child's needs, taking into account the child's strengths, preferences, and interests; and
(C) includes instruction, related services, community experiences, the development of employment and other post-school adult living objectives, and, when appropriate, acquisition of daily living skills and functional vocational evaluation

Table 1. Transition Services as defined by the Individuals with Disabilities Education Improvement Act of 2004 [IDEA, P.L. 108-446, 20 U.S.C. Chapter 34, Section 602 (a) (34)]

Although some STEM occupations only require a high school diploma most require technical training and/or advanced degrees (Carnaevale, Smith, & Melton, 2011; Terrell, 2007). Therefore, transition efforts for most of those interested in STEM should be geared to postsecondary education or training.

Transition activities should also prepare students for integrated employment whether it is sought immediately after high school or after postsecondary education. Employment preparation, such as vocational education, focuses on the development of students' academic and technical knowledge and skills needed to enter their careers or further their education in preparation for career entry. It includes a sequence of organized educational courses that provide integrated academic and vocational instruction designed to improve students' reasoning and problem solving skills, academic knowledge, work attitudes, specific occupational and/or technical skills, and general skills needed for employment. These skills are needed to be successful in STEM -related careers.

Students may need supports available through adult services, to assist them as they work toward independent living or community participation. Providing students with community-based instruction can help them apply academic, social, and/or general work behaviors and skills in natural community settings.

Transition programs and services must be individualized and based on planning that addresses students' post-school goals. Students should be actively engaged in the development of their transition programs and be taught how to direct the planning for their own futures.

Transition planning should reflect the needs, interests, and preferences of students. The input of parents or guardians should also be actively sought and encouraged during transition planning. Parents or guardians might need to be provided training opportunities and access to resources so they can be better able to assist and support their child during the transition process.

Transition planning is an on-going process, but is collectively discussed and drafted into an Individualized Education Program (IEP) annually by students' IEP teams. Table 2 presents sample IEP goals that are consistent with preparing students with disabilities for STEM careers. IEP teams are comprised of the student, his or her parent or guardian, a general education teacher, a special education teacher, a local education agency representative who has the authority to commit agency resources, a person who is knowledgeable and can interpret instructional implications of assessments and others who have knowledge or expertise about the student, including related service personnel (IDEA, 2004). During these meetings assessment results are discussed and students' strengths and needs are identified. A program of study is also determined.

A transition-focused program of study includes courses, experiences, and curriculum designed to develop students' academic and functional achievement, access to post-secondary education or training, and/or attainment of employment and independence. Planning a rigorous course of study that includes as much exposure to math and science as possible is important for those who desire to enter STEM fields (Dunn, Rabren, Taylor, & Dotson, 2012; Terrell, 2007).

The transition planning process also requires the selection of services to help students achieve their post-school goals. These services include instruction that includes access to general education curriculum with specific instruction, services, and/or supports as needed to meet the needs of each student according to his or her disability.

Table 2. Sample IEP

<table>
<tr><td colspan="10">INDIVIDUALIZED EDUCATION PROGRAM</td></tr>
<tr><td colspan="2">STUDENT'S NAME</td><td colspan="8">John Michael Rogers</td></tr>
<tr><td>DOB</td><td>08-06-1998</td><td>SCHOOL YEAR</td><td>2013</td><td>-</td><td>2014</td><td>GRADE</td><td>10</td><td>-</td><td>11</td></tr>
<tr><td colspan="3">IEP INITIATION /DURATION DATES</td><td colspan="2">FROM</td><td>09-15-2013</td><td>TO</td><td colspan="3">09-14-2014</td></tr>
<tr><td colspan="10">STUDENT PROFILE</td></tr>
<tr><td colspan="10">John's academic strengths lie in the areas of reading comprehension and vocabulary development. He currently has an 82 average in English, 75 average in physical science, and 90 average in history. John's social skills are very well developed and he shows a maturity level above those of his peers. He interacts well with younger children and can maintain engaging conversations with adults. He has worked odd jobs during the summer such as cutting grass and washing cars.

John's Transition Planning Inventory results (Home, School, Student forms) indicate that he has strengths in employment, further education/training, daily living, leisure activities, community participation, and health. He scored lower in the areas of self-determination and communication.

John is a member of the school's baseball team. He also spends much of his free time engaged in outdoor sports such as hunting, fishing, and camping. His indoor activities include watching T.V. (e.g., military and science shows and sports). He has attended Explore Science summer camp for the past 3 years at the local community college. He especially enjoyed the small engine and computer sessions. He can use Word, Excel, and is beginning to learn CAD software.

He currently receives math instruction in all general curriculum classes with support from a co-teacher in an inclusive Algebra IA class. John needs to improve his math skills, study habits, and organizational skills. Results of teacher input and grades show that John has difficulties with learning new math concepts. He also needs additional time to solve mathematical problems. His difficulties with study and test taking skills are affecting his progression toward high school graduation and, therefore, have a direct impact on his postsecondary transition goal to attend college.</td></tr>
<tr><td colspan="10">POSTSECONDARY GOALS</td></tr>
<tr><td colspan="10">EDUCATION/TRAINING: After graduating from high school, John will attend a college or university to study mechanical engineering.</td></tr>
<tr><td colspan="10">EMPLOYMENT: After graduating college, John will work in the mechanical engineering field.</td></tr>
<tr><td colspan="10">ANNUAL GOALS</td></tr>
<tr><td colspan="10">By the end of the 2013-14 academic year, John will use study and test taking skills as measured by his use of his study notebook to earn a "B" or better in his math class.</td></tr>
<tr><td colspan="10">INDIVIDUALIZED EDUCATION PROGRAM</td></tr>
<tr><td colspan="10">By May 2014, John will identify no less than two of his strengths and two weaknesses in math.</td></tr>
<tr><td colspan="10">By the end of the 4th nine weeks, John will advocate for himself by requesting as needed and in advance, additional time on math tests.</td></tr>
</table>

Students may also need related services (e.g., vocational rehabilitation, occupational therapy, orientation and mobility training) to help them benefit from their educational programs and assist them in the attainment of their post school goals.

In summary, transition is about preparation for life. It includes the services and programs that are legislatively required to assist those with disabilities who would otherwise have limiting and less than desirable post school outcomes. For those students with disabilities interested in STEM areas, transition services can be the *key* that unlocks the door to exciting and rewarding careers.

TRANSITIONING TO STEM

The previous National Science Education Standards defined equity as "access to skilled professional teachers, adequate classroom time, a rich array of learning materials, accommodating work spaces, and access to the resources of the community surrounding the schools" (p.2). Unfortunately, equity in access for students from traditionally underrepresented groups (Ferreira, 2002), particularly students with disabilities, has yet to be realized. Students with disabilities lag behind their counterparts without disabilities in STEM degree programs and careers and there are few organizations that recruit students with disabilities into STEM programs and careers (Martin, Stumbo, Martin, Collins, Hedrick, Nordstrom, & Peterson, 2011). People with disabilities must be able to successfully navigate various barriers relative to pursuing degree programs and careers in STEM areas (Martin, et al., 2011).

One of the biggest barriers is that of low expectations for students with disabilities. There are a number of stereotypes and stigmas attached to students with disabilities and often times teachers have their own preconceived notions of who can be successful in STEM areas. Consequently, teachers, counselors, and even parents can pass on "hidden messages" to students with disabilities which can inadvertently discourage them from pursuing STEM degrees or continuing in critical gatekeeping courses in precollege years that would prepare them to pursue a degree and careers in STEM related fields (AccessSTEM, 2007; Cobb & Alwell, 2007).

The inability to access science content and participate in science and math experiences is another critical barrier to the participation of students with disabilities in STEM. Access to content and experiences can occur through informal exposure to science education experiences and formal classes. Without access to these experiences early on in elementary school, students may never develop an interest in STEM fields (Moon, Todd, Morton, & Ivey, 2012). Additionally, even though students with disabilities might be enrolled in STEM-related classes, they might not get the true benefits of participation in the classes because many teachers do not know how to adapt instruction and materials so that the content is easier to understand and remember. Further, many teachers do not know how to provide appropriate accommodations (e.g., taking exam in smaller units). These barriers must be considered as programs are designed to increase the numbers and success of students with disabilities in STEM.

Unfortunately, the literature on strategies to increase the participation of persons with disabilities is scant at best, and there are very few programs designed specifically to recruit and retain students with disabilities in STEM areas (Fichten, Asuncion, Barile, Robillard, Fossey, and Lamb, 2003).However, the research-base in transition education that has accrued can be used as a foundation for developing programs that effectively transition high school students with disabilities into STEM careers.

Such programs must provide individually-designed programs and services that include both academics and experiences outside the traditional classroom that promote students' (a) interest and motivation in STEM, (b) confidence in STEM, (c) deep understanding of mathematics and science, and (d) belief that persons with disabilities can be successful in STEM careers (Dunn, et al., 2012; Lee, 2011). Consideration of the legal requirements previously discussed and the research on evidence-based practice in transition and STEM suggest the following program elements should be considered when transitioning students with disabilities into STEM careers: (a) individualized and active student involvement in planning, (b) participation in general education, (c) teaching strategies and accommodations, (d) self-determination, (e) technology, (f) career awareness and career/technical education, (g) experiential development, (h) mentoring, (i) collaboration, and (j) postsecondary connecting activities (Burgstahler, 2011; IDEA 2004; National Secondary Transition Technical Assistance Center, 2013.)

Individualized and Active Student Involvement in Planning

Active student (and family) involvement and individualized planning are recognized as important practices (Bassett & Li, 2008; Morningstar, Wehmeyer, & Dove, 2008). Planning should focus on "the wants and needs of individual youth with disabilities and their families as opposed to simply placing youth with disabilities into available ... programs" (Kochar-Bryant & Greene, 2009, p 221). One of the first steps in this process is having students identify their postsecondary goals. Student goal setting is important as research has shown that students who set college-going goals early have a greater likelihood of attending post-secondary education (Cobb & Alwell, 2007). Goals also provide direction to the development and implementation of a program of study and experiences that help students develop the skills necessary for achieving their goals, in this case, a STEM career. Student-centered planning sessions and student-led conferences (e.g., student-directed IEP, MAPs) are examples of strategies that can be used in school settings (Bassett & Li, 2008) to assist students in identifying interests, goals, and needs (see Table 3 for a list of available materials to support active student involvement in IEPs and self-determination), which form the basis for planning. Students should also be active in the selection of other experiences that will equip them with the foundation for a STEM career (e.g., career exploration, service learning, involvement in work experiences, and connecting activities such as job fairs and special workshops).

Participation in General Education

Participation in general education is critical for all students. It predicts greater success in employment, independent living, and postsecondary education (NSTTAC, 2013). It is particularly relevant for students interested in STEM because some postsecondary education and training is a requirement of the majority of STEMjobs (Carnevale, et al., 2011). Consequently, preparation for postsecondary education or training is vital for the vast majority of students interested in STEM. This preparation is best accomplished through enrollment in general education classes.

Table 3. Resources for Person-Centered Planning, Student Involvement in Transition Planning, and Self-Determination

Material	Summary
Person-Centered Planning http://transitioncoalition.org/transition/section.php?pageId=83	A mini module that provides an overview of person-centered planning techniques that can be used in transition planning.
Person-Centered Planning: A Tool for Transition http://ncset.org/publications/viewdesc.asp?id=1431	An overview of person-centered planning for parents. Includes helpful resources.
Goal Setting Fact Sheet http://www.dcdt.org/wp-content/uploads/2011/09/DCDT_Fact_Sheet__Student_involvement_IEP_Process.pdf	A list of steps to teach goal-setting for parents and professionals. A direct and structured approach.
Youthhood.org http://www.youthhood.org/	An interactive website for students. Gets youth to start thinking about what they want to do for the rest of their lives.
A Student's Guide to the IEP http://nichcy.org/publications#lets	A student workbook and audio program that includes students with disabilities talking about participating in the development of their own IEPs.
My Future My Plan http://www.ncset.org/publications/mfmp.asp	A transition planning resource and video, discussion guide and planning and resource books.
Relish is for More than Hotdogs http://nichcy.org/publications#lets	A student guide and audio program of people with disabilities who pursued dreams.
Youth Fact Sheets http://www.waisman.wisc.edu/hrtw/YFS.pdf	Planning for technical and 4-year college and how to be involved in IEP.
Choice Maker Self-Determination Curriculum www.sopriswest.com	Six instructional packages that teach students self-determination skills needed to be successful in school and adult life.
Whose Future Is It Anyway? http://www.ou.edu/content/education/centers-and-partnerships/zarrow/self-determination-education-materials/whos-future-is-it-anyway.html	A package of lessons to help prepare students for IEP meetings and gain self-determination skills. Teachers serve as a coach in the process.
Steps to Self-Determination www.proedinc.com	A curriculum to help students gain knowledge and skills they need to increase their ability to achieve their goals. Teacher-directed.
Self-Advocacy Strategy http://www.edgeenterprisesinc.com/index.php	*A strategy* students learn to use for preparing for and participating in any type of conference, including education and transition planning conferences.

Moon et al. (2012) cautioned against students with disabilities learning STEM content in special education classes "that do not prepare them for the rigors of university education in STEM fields." (p.10). For students interested in pursuing STEM, the actual classes they take in general education is of utmost importance. Students must take the classes that will help them develop content critical to college success. They must take math and science classes beyond the minimum graduation requirements. The level of courses is also a critical consideration. Research has demonstrated a relationship between successfully completing high school physics and earning a bachelor's degree in STEM fields (Tyson, Lee, Borman, and Hanson, 2007). Additionally, after an analysis of rigorous high schools, American College Testing (ACT) emphasized it is better to take algebra II over and above the more common algebra I and geometry and chemistry over and above the more common biology. Those students who took algebra II and chemistry were more likely to meet or exceed ACT College Readiness Benchmarks (ACT, 2007).

Teaching Strategies and Accommodations

Enrollment in general education STEM classes alone is not sufficient. The instructional strategies and techniques used those classes must make the content accessible to all learners in the class. Universal design and accommodations are two strategies that have been found to be beneficial for increasing the participation and success of students with disabilities in STEM (Burgstahler, 2013). Universal Design for Learning (UDL) is a relatively new concept that consists of a "set of principles for curriculum development that give all individuals equal opportunities to learn" (CAST, n.d.). It encourages teachers to present their material using multiple formats and to allow students to explore and acquire information in various ways (Rao, Dowrick, Yuen, & Boisvert, 2009). UDL also encourages teachers to allow students to showcase their learning in various formats (Rao et al., 2009). This helps ensure teachers have addressed the needs of various types of learners within their classroom. Principles of universal design can be applied to an array of instructional elements including the use of lectures, fieldwork, group work, web-based instruction, printed material, demonstrations, lab work, assessments, and projects (Scott, McGuire, and Shaw, 2003).

Specific techniques and strategies for working with students with different kinds of disabilities are found in chapters 2-12 of this book. However, in general, STEM-focused learning should include hands-on experiments that incorporate lab rotations, STEM competitions designed to motivate students, backyard experiential learning, collaborative gaming that is team-based, and STEM mentors who can supplement the knowledge of the teacher (Kelso, 2011).

Even though employing UDL benefits all learners, some students will need specific accommodations to meet their individual needs (Burgstahler, 2013). A student who is deaf, for example, might need an interpreter and a student with a dexterity or visual impairment might need specialized lab equipment to meet unique individual needs. Unfortunately, some teachers are reluctant to use accommodations, especially in higher level classes, because they believe it gives students an unfair advantage over their peers without disabilities. Teachers need to be educated about the differences between accommodations (i.e., meeting the same standard) and modifications (i.e., changing the standard) and the legitimate need for accommodations. At the same time, students with disabilities need to build a case for their use of accommodations from multiple sources (Brinkerhoff & Banerjee, 2007). One of the key predictors of persistence in STEM is knowing which accommodations work and being able to communicate that to others (Summers, 2008). Students need to practice this while still in school with the support of teachers.

A document that can assist students in identifying and understanding their strengths and accommodation needs is the Summary of Performance (SOP). The SOP is a report that summarizes a student's academic achievements and performance and includes accommodations. It must be completed during the students' final year in high school, although it is helpful to develop it earlier and update as needed (http://www.tknlyouth.org/itp.html). The information included in the SOP can help establish eligibility for reasonable accommodations.

Self-Determination

Self-determination is a key aspect towards the pursuit of STEM degrees and careers. As well, it is considered a critical survival skill for college (Kochhar-Bryant, Bassett, and Webb, 2009). *The American Heritage Dictionary* online (2013) defines self-determination as the "...determination of one's own fate or course of action without compulsion; free will." Although it is important to apply and learn these skills in school, it is more important apply these skills outside of school. Once students leave the safe world of public school, where special services teachers and parents have been their main source of advocacy, students with disabilities will have to learn to depend on themselves. Skills associated with self-determination include: (a) choice making; (b) decision-making; (c) problem-solving; (d) independent living (risk taking and safety skills); (e) goal setting and attainment; (f) self-observation, evaluation, and reinforcement; (g) self-instruction, self-understanding, self-advocacy, and leadership; (h) positive self-efficacy and outcome expectancy; (i) internal locus of control; and (j) self-awareness (Wood, Karvonen, Test, Browder, & Algozzine, 2004, p. 10). These skills are all critical to post school success. For example, recent research has shown that students who have high self-determination skills are more likely to be successful in the transition from high school to post-secondary/employment (Berry, Ward, & Caplan, 2012; Morningstar, Frey, Noonan, Ng, Clavenna-Deane, Graves, & Williams-Diehm, 2010). Further, the U.S. Department of Education, Office of Civil Rights (2011) underscored the need for students with disabilities to be able to identify their disability and understand the functional limitations of it. Educators need to help students develop skills associated with self-determination. An example of self-determination skills development would be learning how to develop goals during the IEP planning process. Teachers also can implement curricula specifically designed to address these skills (see Table 3).

Teachers can further encourage the development of these skills through their use of instructional strategies and environmental considerations (e.g., opportunities for choice, attribution retraining, modeling, behavioral strategies, cooperative learning, coaching) (Field, 2008).

Technology

Over the past ten years, computer and internet technology has slowly improved the way students with disabilities participate and perform within the regular education classroom (Allsopp, McHatton, & Cranston-Gringas, 2009). As a result of federal mandates (e.g., No Child Left Behind Act of 2003, Individuals with Disabilities Education Act of 2004) as well as the development of UDL, more teachers are searching for ways to incorporate innovative and instructional technology into their curriculum (Puckett, Judge, & Brozo, 2009). Computer and internet technology provides a useful tool to support these goals.

Specifically, the creation of educational internet websites has exploded (Allsopp, et al., 2009). Many websites have been created for teachers to use with their students to enhance and support student learning in both the general and special education classroom. Internet tools have several benefits. Because of their visual, technological, and interactive assets they can provide greater motivation for students with disabilities, allow for additional practice time, and give students the opportunity to review previously taught material at their leisure and pace to sustain their newly gained knowledge. Many teachers feel these newer, alternative forms of instruction offer a way to "equalize" the playing field for struggling learners and those who live in more rural areas (Smith & Okolo, 2010). These innovations in instructional technology allow teachers the opportunity to expand the methods in which they presently present information to students with disabilities (Elder-Hinshaw, Manset-Williamson, Nelson, & Dunn, 2006). Table 4 lists websites teachers can use in their classroom to support students in STEM classes. They have been broken down by subject area for easier access. Students must also have access to technology that helps level the playing field and permits students with disabilities to function like those without (e.g., use of a talking calculator for a student who transposes numbers) (American Association for the Advancement of Science, n.d.). In some instances students will need to be taught to use technology that can support them in educational and work environments. For example, they might need to learn how to use modified keyboards or head- or voice-controlled hands-free options as well as software (e.g., word prediction).

Table 4. Technological STEM Tools

Math	Website	Description
NCTM Illuminations	http://illuminations.nctm.org/	Contains resources for teaching mathematics. Includes lessons and activities grades K-12 (e.g. Conic Section Explorer).
SAS Curriculum Pathways	http://www.sascurriculumpathways.com/ portal/	Contains videos, practice problems and online quizzes. Allows you to explore transformations. Has an entire online Algebra I course and textbook with materials.
GeoGebra	http://www.geogebra.org/cms/en/	32,000 activities from kindergarten to university level. Interactive geometry, algebra, statistics, and calculus software.
Quizlet	http://quizlet.com/	Flashcards to study, quiz, and spell vocabulary words. Students can create their own for free or teachers can create flashcards for students to study for class.
Cool Math	http://coolmath.com/	Math review for ages 13 and up (begins with pre-algebra). Re-teaches skills taught in the general education classroom.

Math	Website	Description
Wabbit Graphic Calculator	http://wabbit.codeplex.com	Free downloadable graphic calculator.
DudeFree	http://www.dudefree.com/	Teacher and student worksheets (e.g. unit circle and graphic calculator activities).
IXL	http://www.ixl.com/	Provides many practice problems with detailed explanations to justify answers. K-8 math problems plus algebra and geometry.
Kuta Software	http://kutasoftware.com/	Test and worksheet generators for math teachers.
Web Math	http://www.webmath.com/	Provides answers to specific math questions typed into the program.
Science	**Website**	**Description**
Chem Fiesta	www.chemfiesta.com/	Provides answers to specific questions related to chemistry with practice worksheets, chemistry vocabulary, educational links, questions answered, and an online textbook is in progress.
Geology	http://geology.com/	News and information about geology.
Science Daily	http://www.sciencedaily.com/	Science news and information.
US Geological Survey	http://www.usgs.gov/	Climate, ecosystems, environmental health, hazards, etc
NeoK12	http://www.neok12.com/	Large collection of free, online, educational videos, lessons, quizzes, games and puzzles.
Know Plants	http://knowplants.org/Know_Plants/Home.html	Repository for on-line resources that can be used for plant science education or outreach.
Anatomy Arcade	http://anatomyarcade.com/	Anatomy Arcade makes basic human anatomy come alive through free interactive and videos. Useful for the novice teenager in the classroom to professionals.
Engineering	**Website**	**Description**
Lesson Planet	http://www.lessonplanet.com/search?keywords=engineeringandmedia=lesson	Engineering lesson plans for teachers that coordinate with state and common core standards.
The Physics Classroom	http://www.physicsclassroom.com/	The Physics Classroom is an online, free physics website developed primarily for high school physics students and teachers. The website features a variety of sections intended to support both teachers and students in the tasks of learning and teaching physics.
STEM	**Website**	**Description**
Brightstorm	http://www.brightstorm.com/	Over 2000 math, 600 science and English videos free. Paid subscription for Test Prep videos (ACT, SAT, PSAT, and Advanced Placement).
Hippocampus	http://www.hippocampus.org/	Videos in math, algebra, geometry, calculus, statistics, probability as well as other subjects.
Khan Academy	https://www.khanacademy.org/	Math, algebra, geometry, calculus, and trigonometry and well as science, economics, humanities, and test prep videos.
Scribblar	http://www.scribblar.com/	Free online interactive whiteboard. Online training and tutoring. Students can ask questions they would not ask in class.
Quia (pronounced Kia)		Join and create 16 types of activities (e.g. matching, flashcards, concentration, word search, battleship games, hangman, jumbled words, scavenger hunt, and close procedures). You can

Table 4. (Continued)

STEM	Website	Description
		also create various types of quizzes like multiple choice, matching, fill in the blank, true false, essay, and popup. You can create classes and track quiz scores.
Socrative	http://www.socrative.com/	Socrative is a student response system that empowers teachers to engage their classrooms through a series of educational exercises and games via smart phones, laptops, and tablets. Teachers login through their device and select an activity which controls the flow of questions and games. Students login with their device and interact in real-time with the content. Student responses are visually represented for multiple choice, true/false and short answer questions. For pre-planned activities a teacher can view reports online as a Google spreadsheet or as an emailed Excel file.

Career Awareness and Career/Technical Education

Repetto and Andrews (2012) underscored the importance of a curriculum that includes career development and career/ technical education for middle and high school students with disabilities.

These programs support students in identifying and reaching their postsecondary transition goals and provide career-related work experiences. A commonly used model of career development in special education is the Life-Centered Career Education Model (Brolin & Loyd, 2004). This model identifies four stages of career development. *Career awareness* usually starts in the elementary grades; students learn about different career roles (e.g., employee, friend, parent) and tasks (e.g., teachers help children learn, friends help each other). This is a time when students could be introduced to the wide range of careers available in the STEM areas. *Career exploration* begins in middle school. At this time students begin to compare tasks of various careers to their personal likes and dislikes. Many schools routinely provide career interest assessments to assist students in identifying their areas of academic strength and weakness related to college and workplace success and considering possible career options (e.g., ACT Explore). Participation in a variety of STEM-related activities discussed in the next section is a good way to increase students' career knowledge and confidence. Research indicates that such experiences play an important role in career development (Lam, Doverspike, Zhao, Zhe, & Menzemer, 2008; Malian, 2007). Lam et al. (2008) noted that early experiences "shape self-efficacy, beliefs, and outcome expectations, which in turn affect the formation of vocational interests, which subsequently influence occupational goals, choice actions and performance attainments" (p.22).

Career preparation occurs in high school and is the time when students choose an initial career path. There should be a direct relationship between the courses students take and their postsecondary goals. As noted previously, students interested in STEM careers, must select those academic and career/technical courses that will provide them with knowledge, skills, and experiences to prepare them for postsecondary education or training; further their interest and motivation in STEM; and develop a deep understanding of mathematics, science, and other content and how to apply knowledge and skills. Career and Technical Education (CTE) has expanded the traditional outcome for vocational education of entry-level jobs that require fewer than a 4-year baccalaureate program to a career pathway (Repetto and Andrews, 2012). Students in CTE programs select one of 16 career clusters (i.e., a grouping of many occupations with similar skills and characteristics). One of the 16 career clusters is *Science, Technology, Engineering and Mathematics*. Other clusters, like *Health Science* and *Information Technology*, can also help to build STEM competencies, experiences, and interest. The final stage of career development is *career assimilation*. This refers to the actual transition from high school into postsecondary settings (e.g., university, community college, technical college, employment).

Internships and other work-based learning also provide valuable STEM experiences. A significant number of students with disabilities have limited or no work experience and have difficulty understanding the impact their disability has on selecting a career. Therefore, internships or other types of work-based programs can "provide a critical link between the academic setting and the work environment to enable all students, particularly students with disabilities, to apply their knowledge and determine the appropriate work environments that best match their skills and abilities" (Briel & Getzel, 2001, p.9).

Experiential Development

Providing early and varied STEM experiences is important to increasing interest in STEM as well as preparing students to succeed in STEM programs. Many elementary, middle and high school students have a negative attitude toward science and math. They might not see them as important or think the classes are too hard. One way to address this is through early and continued exposure to age-appropriate STEM opportunities. Examples of such opportunities include STEM afterschool and summer opportunities, special programs, as well as extracurricular activities.. These types of programs supplement traditional academics with authentic and hands-on experiences that are acknowledged as effective strategies for increasing diversity in STEM (Liston, Peterson, & Ragan, 2007; Malian, 2007). These experiences play an important role in increasing students' interest, confidence, career knowledge, motivation, access, and preparedness for STEM (Afterschool Alliance, 2011; Lam, et al., 2008; Mailin, 2007; Mastorpieri & Scruggs, 1992). As well, participation in such programs has been linked to higher rates of graduation from high school and pursuit of college and intention of majoring in STEM (Afterschool Alliance, 2011).

Supplemental programs and experiences come in many shapes and forms. They can be in person or virtually through on-line exchanges. They can be formal or informal. Or, they can occur during the traditional school day or afterschool, in the evenings, on weekends and during the summer. They can supplement the school curriculum (i.e., aligned with school curriculum) or complement the school curriculum (i.e., a greater focus on exploration and development of inquiry skills) (National Partnership for After School Education, n.d.).

After-school programs have great potential for facilitating STEM learning, increasing self-esteem, and promoting positive attitudes for students with disabilities. These programs "can provide youth with a safe and supportive adult-supervised environment and offer them various growth-enhancing opportunities, including activities and experiences that promote academic, personal, social and recreational development" (Durlak & Weissberg, 2007, p. 5). They can provide a level of individualized attention that many students with disabilities might not receive during the school day and give them a better chance to reach their potential (Afterschool Alliance, 2004). An increasing number of these programs have a STEM focus. An example is the High School/High Tech program in Sarasota, Florida. This program is a partnership with Sarasota-based Dolphin Aviation, providing students with disabilities the opportunity to fly (Afterschool Alliance, 2004). STEM benefits have been demonstrated through participation in such national programs as 4-H in which 71% of 4-H science participants indicated that science was one of their favorite subjects and 59% expressed interest in a job in science (Krishnamurthi, Ottinger, & Topol, 2013).

The benefits of participation in extracurricular activities are many. Not only does it promote socialization and the development of leadership skills, but it also provides another way to provide STEM-related experiences. Through participation in these activities students can engage in extensions of their academic curriculum, increase confidence in their STEM abilities, establish a network of other students with shared interests, and explore career paths. Some may even earn recognition through scholarships and awards. School clubs in areas such as chess, ecology or robotics are one type of extracurricular activity. There are a range of competitions in which students can participate; some are team-based while others are for individuals. Examples include First Robotics, Math and Science Olympiads, National Science Bowl, NASA Space Settlement Contest, MATHCOUNTS, BEST Robotics, Real World Design Challenge, National STEM Video Game Challenge, Microsoft Imagine Cup, and Google Science Fair. (See http://stemology.wordpress.com/stem-competitions/ for additional competitions). STEM development can also be promoted through various service learning activities in which students work together with STEM industry advisors, teachers, higher education faculty and others to meet STEM-related community needs.

Other kinds of specialized programs in which students can participate are charter schools, private schools, and magnet schools that specialize in STEM. More and more of these programs target underrepresented groups, which includes students with disabilities. Precollege STEM programs are also offered from many universities. These programs provide experiences in a wide array of STEM areas, with many also focusing on orientation to college. The National Science Foundation's Research in Disabilities Education Programs funds Alliance projects that are targeted to people with disabilities in STEM. Many of these projects host after-school, summer, weekend, or even residential programs for students with disabilities interested in STEM.

Mentoring

Mentors and role models can bevaluable educational resources. Powers, Sowers, and Keller (n.d.) recently conducted a randomized study on the impact of STEM mentors on high school students with disabilities. A key finding was that mentoring seemed to be a good strategy for imparting practical knowledge about STEM fields and what steps to take to pursue them, which impacted students' confidence related to the information. As well, for most of the students with disabilities, it did not make a difference whether the STEM mentor had a disability or not. And, finally, older students (16-17) had greater gains than younger students (14-15). The researchers thought that maybe younger students needed more "fun" STEM activities. It could be that the older students were more interested in specifics such as how to apply to college and what college is like.

Mentoring can occur in many ways. It can occur via one-on-one or small group and through face-to-face meetings, e-mail exchanges, Skype, telephone conversations, or other forms of communication (Sword and Hill, 2002). Successful mentoring programs for students with disabilities include mentor training, flexibility, and opportunities for relationship building and social role development (Stumbo et al., 2010-2011). Stumbo et al. noted that many effective mentoring programs provide additional STEM-related activities such as campus tours, summer camps, and community events. When local mentoring programs are not available, On-line mentoring programs such as those sponsored by AccessComputing, AccessSTEM, Institute for Human Centered Desalign, DO-IT Pals, and CareerConnect, are on option when local mentoring is not available.

Another promising practice for recruiting students into STEM and helping lay the foundation is student learning communities (SLCs), which are networks essential for student engagement in STEM and potentially increase numbers of students with disabilities in the STEM workforce (Izzo, Murray, Priest, & McArrell, 2011). SLCs allow students with disabilities to develop self-advocacy plans, become more aware of their personality and learning styles, and learn more about STEM fields (Izzo, et al., 2011).

Collaboration

When taking into consideration the many program components necessary to successfully transition a student into a STEM program, it should be evident the critical role collaboration plays. The successful transition of students is a shared responsibility. Unfortunately, the many stakeholders (e.g., students, parents, general education content teachers, special education teachers, counselors, mentors, businesses, extracurricular activity sponsors) involved in developing and providing secondary students with disabilities a strong STEM foundation likely have different experiences, levels of knowledge, and perspectives on students with disabilities in STEM. Intentional efforts across this range of individuals must be made to coordinate appropriate academic content; effective modes of providing instruction, necessary accommodations, and individualized supports; and other necessary programs and experiences (e.g., mentoring, internships, extracurricular activities) to meet the unique needs of the individual.

Postsecondary Connecting Activities

A final important planning consideration is helping students connect with supports and services from institutions of higher education prior to beginning their postsecondary education. An array of supports and services are available to all students and some will be available only to students with documented disabilities. They include disability service and mentoring programs, advocacy/self-advocacy groups, and internships, all of which have been found to affect the success of students with disabilities in STEM (Summers, 2008). Students must establish their eligibility for some of these services early, preferably before they start school. The way in which services are obtained and provided at the postsecondary level differs significantly from high school. Students must self-identify, request accommodations, and initiate involvement in groups on their own. Therefore, ensuring that students connect with and establish eligibility with programs prior to starting school is important to their postsecondary success.

Conclusion

In an increasingly global and technological society it is critical that educators provide experiences that promote scientific literacy, provide as many relevant science learning experiences as possible, and maximize participation of all students in STEM degree programs and careers. Lack of student motivation and interest in science and math learning experiences can often result in their underrepresentation in STEM careers (Russell, 2014). Expectations for "who can do science" must be changed and the status quo must be challenged to encourage students from all backgrounds to participate and succeed at the highest levels in science (Butler, Atwater, & Russell, 2014). This chapter provided strategies for increasing the participation and success of students with disabilities and other special needs in STEM. The foundation for a comprehensive approach to transitioning such students into STEM is the special education transition literature. By implementing practices that promote students' interest and motivation in STEM, development of a deep foundation of STEM content and careers, acquisition of self-determination skills, and collaboration across the diverse range of stakeholders, students should be better prepared to transition to a postsecondary STEM program and ultimately a fulfilling STEM career.

References

Afterschool Alliance. (2004). *Afterschool Alert*, Issue Brief No. 1 Afterschool and students with special needs. Retrieved from http://www.afterschoolalliance.org/ issue_1_ needs.cfm

Afterschool Alliance. (2011). *STEM learning in afterschool: An analysis of impact and outcomes.* Retrieved from http://www.afterschoolalliance.org/STEM-Afterschool-Outcomes.pdf

Allsopp, D. H., McHatton, A., & Cranston-Gingras, A. (2009). Examining perceptions of systematic integration of instructional technology in a teacher education program. Teacher Education & Special Education: *The Journal of the Teacher Education Division of the Council for Exceptional Children,* 32, 337-350.

Alston, R., & Hampton, J. (2000). Science and engineering as viable career choices for students with disabilities: A survey of parents and teachers. *Rehabilitation Counseling Bulletin,* 43(3), 158-164.

Alston, R., Bell, T., & Hampton, J. (2002). Learning disability and career entry into the sciences: A critical analysis of attitudinal factors. *Journal of Career Development*, 28(4), 263-275.

American Association for the Advancement of Science. (n.d.). New career paths for students with disabilities: Opportunities in science, technology, engineering, and mathematics. Washington, DC: Author.

American College Testing. (2007). *Rigor at risk: Reaffirming quality in the high school core curriculum.* Iowa City, IA: Author.

Atwater, M. M., Russell, M., & Butler, M. B. (2014). Introduction: Culture, Equity, and Social Justice for Science Teacher Educators. In *Multicultural Science Education: Preparing teachers for equity and social justice,* (pp. 1-7). Springer, Netherlands

Bassett, D., & Li, J. (2008). Student involvement in transition-related activities. In G. Blalock, J. Patton, P. Kohler, and D. Bassett (Eds.). Transition and students with learning disabilities: Facilitating the movement from school to adult life (2nd ed.) (pp. 51-78). Austin, TX: Hammill Institute on Disabilities.

Benz, M. R., Lindstrom, L., & Yovanoff, P. (2000). *Improving graduation and employment Exceptional Children,* 66, 509–529.

Berry, H., Ward, M., & Caplan, L. (2012). Self-determination and access to postsecondary education in transitioning youths receiving supplemental security income benefits. *Career Development and Transition for Exceptional Individuals*, 35(2), 68-75.

Briel, L., & Getzel, E. (2001). Internships in higher education: Promoting success for students with disabilities. *Disability Studies Quarterly*, 21(1), 1-9.

Brinkerhoff, L., & Banerjee, M. (2007). Misconceptions regarding accommodations on high stakes tests: Recommendations for preparing disability documentation for test takers with learning disabilities. *Learning Disabilities Research and Practice*, 22(4), 246-255.

Brolin, D., & Lloyd, R. (2004). *Career development and transition services: A functional life skills approach* (4th ed.). Upper Saddle River, NJ: Pearson.

Burgstahler, C. (2012). *Universal design of instruction (UDI): Definition, principles, guidelines and examples.* Seattle, Washington: DO-IT Center. Retrieved from http://www.washington.edu/doit/Brochures/Academics/instruction.html

Burgstahler, C. (2011). *AccessSTEM: Progress of teens with disabilities toward STEM Careers.* Seattle, Washington: University of Washington. Retrieved from http://www.washington.edu/doit/Stem/flowchart.html

Burgstahler, C. (2013) Equal access: Universal design of instruction. Seattle: University of Washington, DO-IT. Retrieved from http://www.washington.edu/doit/Brochures/Academics/equal_access_udi.html

Cameto R., Levine, P., & Wagner, M. (2004). *Transition planning for students with disabilities.* Menlo Park, CA. SRI International. Retrieved from www.nlts2.org/reports/2004_11/nlts2_report_2004_11_complete.pdf

Carnevale, A., Smith, N., & Melton, M. (2011). STEM, Science Technology Engineering Mathematics. Georgetown University: Center on Education and the Workforce.

Center for Applied Special Technology (CAST). (n.d.). Transforming education through universal design for learning. Retrieved from http://www.cast.org/index.html

Disability Statistics and Demographics Rehabilitation Research and Training Center.(2011). 2011 Annual Disability Statistics Compendium. Durham, NH: Institute on Disability.

Dunn, C. Rabren, K. S., Taylor, S. L., & Dotson, C. K. (2012). Assisting students with high-incidence disabilities to pursues careers in science, technology, engineering and mathematics. Intervention in School and Clinic, 48, 47-54. doi: 10.1177/1053451212443151

Durlak, J. A., & Weissberg, R. P. (2007). The impact of after-school programs that promote personal and social skills. Chicago, IL: Collaborative for Academic, Social, and Emotional Learning.

Elder-Hinshaw, R., Manset-Williamson, G., Nelson, J. M., & Dunn, M. W. (2006). Engaging older students with reading disabilities: Multimedia inquiry projects supported by reading assistive technology. *Teaching Exceptional Children,* 39(1), 6-11.

Eriksson, L., Welander, J., & Granlund, M. (2007). Participation in everyday school activities for children with and without disabilities. *Journal of Developmental and Physical Disabilities,* 19(5), 485-502.

Fichten, C. S., Asuncion, J. V., Barile, M., Robillard, C., Fossey, M. E., & Lamb, D. (2003). Canadian postsecondary students with disabilities: Where are they? *The Canadian Journal of Higher Education*, 33(3), 71-114.

Field, S., (2008). Self-determination instructional strategies for youth with learning disabilities. In G. Blalock, J. Patton, P. Kohler, and D. Bassett (Eds.). Transition and students with learning disabilities: Facilitating the movement from school to adult life (2nd ed.) (pp. 167-201). Austin, TX: Hammill Institute on Disabilities.

Grossman, C. R. S. (2008). Preparing WIA Youth for the STEM Workforce. Youth work Information Briefs [serial on the Internet]. Retrieved from http://jfs.ohio.gov/owd/WorkforceProf/Youth/Docs/Infobrief35_STEM_Workforce_.pdf

Halpern, A. (1992). Transition: Old wine in new bottles. *Exceptional Children*, 58, 202–211.

Individuals with Disabilities Education Improvement Act of 2004, Public Law 108-446, 118, Stat. 2647.

Izzo, M., Murray, A., Priest, S., & McArrell, B. (2011). Using student learning communities to recruit STEM students with disabilities. *Journal of Postsecondary Education and Disability,* 24(4), 301-316.

Johnson, D. R. (2012). Policy and adolescent transition education. In M. Wehmeyer and K. Webb (Eds.) Handbook of Adolescent Transition Education for Youth with Disabilities (pp. 11-31). New York: Routledge.

Kelso, K. (2011). A strategy paper: Digital STEM learning and the high school student. Center for Digital Education. Retrieved from http://images.centerdigitaled.com/documents/CDE10+STRATEGY+Dell_HighSchool_V.pdf

Kochar-Bryant, C., & Greene, G. (2009). Pathways to successful transition for youth with learning disabilities. Pearson: Columbus, OH.

Kochar-Bryant, C. A., Bassett, D. S., & Webb, K. W. (2009). Transition to postsecondary education for students with disabilities. Thousand Oaks, CA: Corwin Press.

Krishnamurthi, A., Ottinger, R., & Topol, T. STEM Learning in Afterschool and summer programming: An essential strategy for STEM education reform. In T. Peterson (Ed.) Expanding Minds and Opportunities: Leveraging the Power of Afterschool and Summer Learning for Student Success (p. 133-139). Retrieved from http://www.expandinglearning.org/expandingminds/article/stem-learning-afterschool-and-summer-programming-essential-strategy-stem

Lam, P., Doverspike, D., Zhao, J., Zhe, J., & Menzemer, C. (2008). An evaluation of a STEM program for middle school students on learning disability related IEPs. *Journal of STEM Education*, 9(1-2), 21-29.

Lee, A. (2011). Postsecondary science, engineering, and mathematics (STEM) enrollment comparisons for students with and without disabilities. *Career Development for Exceptional Individuals,* 34(2), 72-82.

Liston, C., Peterson, K., & Ragan, V. (2007). Guide to promising practices in information technology education for girls. Boulder, CO: National Center for Women and Information Technology.

Madaus, J. W., Foley, T. E., McGuire, J. M., & Ruban, L. M. (2002). Employment self-disclosure of post-secondary graduates with learning disabilities. *Journal of Learning Disabilities,* 35(4), 364-369.

Malian, I. (2007). MVIP: Math village for inclusive practices: A model to engage all students and teachers in STEM experiences. *Electronic Journal for Inclusive Education,* 7(2). Retrieved from http://www.cehs.wright.edu/resources/publication/ejie/ WinterSpring2011 /index.html.

Markel, K. S., & Barclay, L. A. (2009). Addressing the underemployment of persons with disabilities: Recommendations for expanding organizational social responsibility. *Employee Responsibilities and Rights Journal,* 21, 305-18.

Martin, J. K., Stumbo, N. J., Martin, L. G., Collins, K. D., Hedrick, B. N., Nordstrom, H., & Peterson, M. (2011). Recruitment of students with disabilities: Exploration of science, technology, engineering, and mathematics. *Journal of Postsecondary Education and Disability,* 24(4), 285-299.

Mastropieri, M. A., & Scruggs, T. E. (1992). Science for students with disabilities. *Review of Educational Research,* 62, 377-411.

Moon, N. W., Todd, R. L., Morton, D. L., & Ivey, E. (2012). Accommodating students with disabilities in science, technology, engineering, and mathematics (STEM): Findings from research and practice for middle grades through university education. Atlanta, GA: Center for Assistive Technology and Environmental Access.

Morningstar, M., Frey, B., Noonan, P., Ng, J., Clavenna-Deane, B., Graves, P., Kellems, R., & Williams-Diehm, K., (2010). A preliminary investigation of the relationship of transition preparation and self-determination for students with disabilities in postsecondary educational settings. *Career Development for Exceptional Individuals,* 33(2), 80-94.

Morningstar, M., Wehmeyer, M., & Dove, C. (2008). Role of families in enhancing transition outcomes for youth with learning disabilities. In G. Blalock, J. Patton, P. Kohler, and D. Bassett (Eds.). Transition and students with learning disabilities: Facilitating the movement from school to adult life (2nd ed.) (pp. 79-103). Austin, TX: Hammill Institute on Disabilities.

Nagle, K., Marder, C., & Schiller, E. (2009). Research in disabilities education program evaluation: Study I methods and results. Arlington, VA: SRI International.

National Partnership for After School Education. (n.d.). What is afterschool? Retrieved from http://www.pasesetter.com/aboutPASE/afterschool.html.

National Secondary Transition Technical Assistance Center. (2013). Predictors of post school success. Retrieved from http://www.nsttac.org/sites/default/files/assets/pdf/pdf/ebps/ SummaryOfPredictorsCategories.pdf

Newman, L., Wagner, M., Cameto, R., & Knokey, A. M. (2009).The post-high school outcomes of youth with disabilities up to 4 years after high school: A report of findings from the National Longitudinal Transition Study–2 (NLTS2) (NCSER 2009-3017).Menlo Park, CA: SRI International. Retrieved from www.nlts2.org/reports/2009_04/ nlts2_report_2009_04_complete.pdf

Newman, L., Wagner, M., Knokey, A. M., Marder, C., Nagle, K., Shaver, D., Schwarting, M. (2011). The post-high-school outcomes of young adults with disabilities up to 8 years after high school: A report from the National Longitudinal TransitionStudy-2 (NLTS-2) (NCSER 2011-3005). Menlo Park, CA: SRI International.

Powers, L., Sowers, J., & Keller, T. (n.d.). A randomized study of the impact of STEM mentors on high school students with disabilities. [PowerPoint slides]. Retrieved from https://www.rri.pdx.edu/files/39/stem_mentor_ppt.pdf

Price, L., Gerber, P., & Mulligan, R. (2007). Adults with learning disabilities and the underutilization of the Americans with Disabilities Act. *Remedial and Special Education,* 28(6), 340-344.

Puckett, K., Judge, S., & Brozo, W. (2009). Integrating content area literacy and assistive technology: A teacher development institute. *Southeastern Teacher Education Journal,* 2(2), 27-38.

Rao, K., Dowrick, P. W., Yuel, J., & Bolsvert, P. C., (2009). Writing in a multimedia environment: Pilot outcomes for high school students in special education. *Journal of Special Education Technology*, 24(1), 27-38. Repetto, J. B., and Andrews W. D. (2012). Career development and vocational instruction. In M. Wehmeyer and K. Webb (Eds.). Handbook of Adolescent Transition Education for Youth with Disabilities. (pp. 156-170). New York: Routledge.

Russell, M. L. (2014). Motivation in the Science Classroom: Through a Lens of Equity and Social Justice. In *Multicultural Science Education: Preparing teachers for equity and social justice* (pp. 103-116). Springer Netherlands.

Sanford, C., Newman, L., Wagner, M., Cameto, R., Knokey, A. M., & Shaver, D. (2011). The post-high school outcomes of young adults with disabilities up to 6 years after high school. Key finding from the National Longitudinal Transition Study-2 (NLTS-2) (NCSER 2011-3004). Menlo Park, CA: SRI International. Retrieved from www.nlts2.org/reports.

Scanlon, D., & Mellard, D. (2002). Academic and participation profiles of school-age dropouts with and without disabilities. *Exceptional Children,* 68(2), 239–258.

Schiavelli, M. (2011). STEM Jobs Outlook Strong, but Collaboration Needed to Fill Jobs. Retrieved from http://www.usnews.com/news/blogs/stem-education/2011/11/03/stem-jobs-outlook-strong-but-collaboration-needed-to-fill-jobs

Scott, S. S., McGuire, J. M., & Shaw, S. F. (2003). Universal design for instruction: A new paradigm for adult instruction in postsecondary education. *Remedial and Special Education*, 24(6), 369-380.

Smith, S. J., & Okolo, C. (2010). Response to intervention and evidence-based practices: Where does technology fit? *Learning Disability Quarterly*, 33(4), 257-272

Stumbo, N., Martin, J., Nordstrom, D., Rolfe, T., Burgstahler, S., Whitney, J., & Misquez, E. (2010-2011). Evidence-based practices in mentoring students with disabilities: Four case studies. *Journal of Science Education for Students with Disabilities*, 14(1), 33-54.

Summers, L. (2008). The role of diversity in higher education: Students with disabilities in STEM. Paper presented at EPSCOR Conference, Charleston, WV. Retrieved from http://www.wvresearch.org/wpcontent/uploads/2011/11/summers.pdf

Sword, C., & Hill, K. (2002). Creating mentoring opportunities for youth with disabilities: Issues and suggested strategies. In Issue Brief: Examining current challenges in secondary education and transition (Vol. 1, Issue 4). Minneapolis, MN: National Center on Secondary Education and Transition. Retrieved from http://www.ncset.org/ publications/viewdesc.asp?id=704

Terrell, N. (2007). Science, technology, engineering, and mathematics occupations. Occupational Outlook Quarterly, 1(1) 26-33. Retrieved from http://www.bls.gov/ opub/ooq/2007/spring/art04.pdf.

The American Heritage Dictionary of the English Language (2013). Self-determination. Retrieved from http://ahdictionary.com/word/search.html?q=selfdetermination+ andsubmit.x=29andsubmit.y=22.

Tyson, W., Lee, R., Borman, K., & Hanson, M. (2007). Science, technology, engineering and mathematics (STEM) pathways: High school science and math coursework and postsecondary degree attainment. *Journal of Education for Students Placed at Risk*, 12(3), 243-270.

U.S. Department of Labor, Bureau of Labor Statistics. (2012). "News Release: The Employment Situation—December 2011." Retrieved from http://www.bls.gov/ news.release/pdf/empsit.pdf.

U.S. Department of Education, Office for Civil Rights. (2011). Students With Disabilities Preparing for Postsecondary Education: Know Your Rights and Responsibilities. Washington, D.C.

U.S. Senate Committee on Health, Education, Labor, and Pensions. (July 2012). Unfinished business: Making employment of people with disabilities a national priority. Retrieved from *www.harkin.senate.gov/documents/pdf/500469b49b364.pdf*

Wood, W., Karvonen, M., Test, D., Browder, D., & Algozzine, B. (2004). Promoting student self-determination skills in IEP planning. *Teaching Exceptional Children*, 36(3), 8-16.

EDITOR CONTACT INFORMATION

Dr. Satasha L. Green
New York Institute of Technology
Dean of the School of Education
Northern Boulevard-Wisser Library
P.O. Box 8000
Old Westbury, NY 11568-8000
Tel: 1-516-686-7706
E-mail: slgreen@nyit.edu

INDEX

B

C

F

G

H

I

J

K

L

N

O

Q

R

S

T

U

V

W

Y